| DATE DUE | | | |
|---|---|---|---|
| Aug 10'83 | | | |
| | | | |
| | | | |
| | | | |
| | | | |
| | | | |
| | | | |
| | | | |
| | | | |
| | | | |
| | | | |

# The Hydrogen
# Energy Economy

# Edward M. Dickson
# John W. Ryan
# Marilyn H. Smulyan

The Praeger Special Studies program—
utilizing the most modern and efficient book
production techniques and a selective
worldwide distribution network—makes
available to the academic, government, and
business communities significant, timely
research in U.S. and international economic, social, and political development.

# The Hydrogen Energy Economy

## A Realistic Appraisal of Prospects and Impacts

**Praeger Publishers**   New York   London

PRAEGER SPECIAL STUDIES IN U.S. ECONOMIC, SOCIAL, AND POLITICAL ISSUES

Library of Congress Cataloging in Publication Data

Dickson, Edward M
    The hydrogen energy economy.

    (Praeger special studies in U.S. economic, social, and political issues)
    Includes bibliographies.
    1. Hydrogen as fuel.  I.  Ryan, John W.,
1940-     joint author.  II.  Smulyan, Marilyn H.,
1948-     joint author.  III.  Title.
TP360.D5      338.4'7'66581     76-56807
ISBN 0-275-24290-0

This book was prepared with the support of NSF Grant
No. ERS73-02706. However, any opinions, findings,
conclusions, or recommendations herein are those of
the authors and do not necessarily reflect the views of NSF.

PRAEGER PUBLISHERS
200 Park Avenue, New York, N.Y. 10017, U.S.A.

Published in the United States of America in 1977
by Praeger Publishers, Inc.

789  038  987654321

Printed in the United States of America

This technology assessment is "preliminary" only in the sense that, with the funds available, the full range of societal consequences of the hydrogen economy could not be pursued to their conclusion. Instead, most attention was devoted to those sectors where the study team felt the consequences would be most significant or most widely distributed, or where the nature of the technology and the state of the institutional setting suggest that implementation would be most difficult.

The study was funded by the Research Applied to National Needs (RANN) Program of the National Science Foundation (Grant ERP73-02706). The program managers were, successively, Joseph Coates and Patrick Johnson.

The study team was led by Edward M. Dickson, who also performed most of the analysis; John W. Ryan was responsible for economic implications and is primarily responsible for Chapter 11; Marilyn Smulyan contributed the social analysis portions of Chapters 8, 14, 15, and 16. Leo Weisbecker contributed some early analysis of institutions, especially in the aviation sector, and Thomas Logothetti contributed some preliminary analysis of social consequences. Within the Stanford Research Institute (SRI) the project was supervised first by George D. Hopkins and later by Robert M. Rodden.

This study has benefited greatly from the participation of two team members (Dickson and Ryan) on a concurrent SRI project entitled "Impacts of Synthetic Liquid Fuel Development—Automotive Market," which was also a technology assessment. The latter study concerned the development of liquid fuels (synthetic crude oil and methanol) from coal and oil resources. Although the gradual introduction of these fuels should prove easier to implement than a transition to hydrogen because of their great compatibility with existing systems, the formidable difficulties of even such a limited transition served as a sobering backdrop against which to evaluate the dynamics of a transition to hydrogen.

The study team wishes to acknowledge the expert and patient editing of Phyllis Dorset and the graphics design of L. H. Wu.

# CONTENTS

| | Page |
|---|---|
| PREFACE | v |
| LIST OF TABLES | xiv |
| LIST OF FIGURES | xvii |
| LIST OF ABBREVIATIONS | xx |
| NOTE TO THE READER | xxi |

Chapter

## PART I: MAJOR CONCLUSIONS AND RECOMMENDATIONS

| | |
|---|---|
| Use of Hydrogen | 3 |
| Transition to Hydrogen | 4 |
| The Future of Hydrogen | 9 |
| Recommendations for Research and Development | 10 |

## PART II: TECHNOLOGY ASSESSMENT AND ENERGY IN THE FUTURE

| | | |
|---|---|---|
| 1 | THE CONCEPT OF TECHNOLOGY ASSESSMENT | 17 |
| | References | 19 |
| 2 | ENERGY AND HYDROGEN IN THE FUTURE | 20 |
| | Change Is Inevitable | 20 |
| | Differences Between the Past and the Future | 21 |
| | The Possible Role of Hydrogen | 23 |
| | Advantages of Hydrogen | 23 |
| | The Need for a Technology Assessment | 24 |
| | References | 25 |

Chapter                                                                    Page

3    METHODS OF THE STUDY                                                   28

     Bibliography                                                           30


              PART III:  HYDROGEN TECHNOLOGIES—
                    PRESENT AND PROJECTED


4    HYDROGEN PRODUCTION                                                    35

     Sources of Hydrogen                                                    35
     Hydrogen Production from Fossil Fuels                                  35
     Hydrogen from Water by Electrolysis                                    38
     Closed-Cycle Thermochemical Decomposition of Water                    42
     Mixed Thermal/Electrolytic                                            45
     Hydrogen from Thermonuclear Fusion                                    45
     Comparison of Hydrogen Production Processes                           47
     References                                                            47


5    STORAGE OF HYDROGEN                                                    51

     Introduction                                                          51
     Storage of Gaseous Hydrogen                                           51
       Underground Storage                                                 51
       Pipeline Storage                                                    53
     Storage of Liquid Hydrogen                                            53
     Liquefaction for Storage                                              55
     Storage in the Form of Metal Hydrides                                 57
     Storage in Chemical Compounds                                         61
     References                                                            62


6    DISTRIBUTION OF HYDROGEN                                               64

     Distribution in Gaseous Forms                                         64
     Distribution in Liquid Form                                           68
     Distribution in Metal Hydride Form                                    69
     Summary                                                               70
     References                                                            70


7    END-USES OF HYDROGEN                                                   73

     Versatility of Application                                            73
     Energy Utilities                                                      73
       Gas Utilities                                                       73

| Chapter | | Page |
|---|---|---|
| | Electric Utilities | 78 |
| | Combined Electric and Gas Utilities | 84 |
| | Automotive Applications | 84 |
| | Private Passenger Vehicles | 84 |
| | Fleet Vehicles | 88 |
| | Off-the-Road Vehicles | 88 |
| | Aircraft Applications | 89 |
| | Ship, Train, and Spacecraft Applications | 91 |
| | Military Applications | 94 |
| | Chemical Applications | 95 |
| | Ammonia Synthesis | 96 |
| | Coal Gasification or Liquefaction | 97 |
| | Minor Chemical Uses | 97 |
| | Chemical Reduction | 98 |
| | Cryogenic Applications | 99 |
| | References | 100 |
| 8 | HYDROGEN SAFETY | 106 |
| | Complexity of the Safety Question | 106 |
| | Properties of Hydrogen | 106 |
| | Hydrogen Behavior | 108 |
| | Experience in the Space Program and the Hindenburg | 111 |
| | Special Hazards | 113 |
| | Summary of Physical and Chemical Aspects of Safety | 114 |
| | Perceptions of Safety Hazards—A Key to Policy | 114 |
| | Voluntary Versus Involuntary Exposure to Hazards | 115 |
| | References | 116 |

PART IV: COMPETING AND COMPLEMENTING
TECHNOLOGIES: PRESENT AND PROJECTED

| 9 | ENERGY CARRIER, DISTRIBUTION, AND STORAGE ALTERNATIVES TO HYDROGEN | 121 |
|---|---|---|
| | Energy Carrier | 121 |
| | Energy Distribution | 124 |
| | Energy Storage | 127 |
| | References | 128 |

Chapter                                                        Page

10    ENERGY END-USE ALTERNATIVES TO HYDROGEN        129

      Energy Utilities to Consumers                      129
        Gas                                              129
        Electricity                                      130
      Transportation                                     130
        Private Automobiles                              130
        Fleet Vehicles                                   132
        Ships and Trains                                 134
        Commercial Aircraft                              135
      Iron Ore Reduction                                 135
      Summary                                            136
      References                                         136


                PART V:  ECONOMICS OF HYDROGEN

11    HYDROGEN COSTS AND ECONOMIC RELATIONSHIPS
      TO OTHER FUELS                                     141

      Development of Cost Estimates                       141
      Hydrogen Production Costs                           143
        Hydrogen from Coal                                144
        Hydrogen from Electrolysis of Water               146
        Hydrogen from Thermochemical Processes            147
        Summary Comparison of Hydrogen Production Costs    150
      Hydrogen Transport Costs                            151
        Gaseous Hydrogen                                  151
        Liquid Hydrogen                                   154
      Hydrogen Storage Costs                              156
        Gaseous Hydrogen                                  156
        Liquid Hydrogen                                   157
      Costs of Other Fuels Compared to Hydrogen Systems   158
      Governmental Actions in Energy Pricing              160
      References                                          161


            PART VI:  TRANSITION TO A HYDROGEN ECONOMY

12    SYSTEMS DESCRIPTION                                 165

      The Building-Block Concept                          165
      Hydrogen Economy Building Blocks                    166

Chapter                                                          Page

13    TRANSITION SCENARIOS                                        170

      Important Noneconomic Factors                              170
      Scenarios                                                  171
      Summary and Discussion                                     178


                  PART VII:  CONSEQUENCES OF
                     A HYDROGEN ECONOMY

14    IMPACTS OF HYDROGEN-FUELED PRIVATE AND
      FLEET AUTOMOTIVE VEHICLES                                   183

      Private Automobiles                                        183
        The System and the Stakeholders                          183
        The Dilemma                                              187
        Transition Strategies                                    188
        Air Quality Implications                                 189
      Social Impacts                                             194
        Safety Considerations                                    194
        Specialization Required in Production and Mainte-
          nance                                                  195
        Decreased Independence and Self-Reliance                 196
        Reduction in Air Pollution                               197
        Independence from Foreign Energy Sources                 197
        Impacts of Hydrogen Filling Stations                     198
      Economic Impacts                                           199
      Resource Utilization                                       203
      Fleet Vehicles                                             205
        Decisions to Deploy                                      205
        Impacts                                                  206
      Summary                                                    209
      References                                                 209

15    COMMERCIAL AVIATION                                         213

      The System and the Stakeholders                            213
        The Airlines                                             214
        Aircraft Manufacturers                                   217
        Airport Operators                                        219
        Fuel Producers and Distributors                          220
        Government Regulatory Agencies                           221
      Transition to Hydrogen                                     227

Chapter                                                                Page

    Impacts                                                          232
      Economic                                                       232
      Environmental                                                  237
      Resource Utilization                                           241
      Social                                                         243
      Social Effects of Safety                                       244
      Technology                                                     246
    Governmental Role                                                247
    Summary                                                          250
    References                                                       250

16    UTILITIES                                                     253

    Introduction                                                     253
    Gas Utilities                                                    255
      Structure and Stakeholders                                     255
      Consumers                                                      257
      Electric Power Generation                                      258
      Industrial Use                                                 260
      Barriers and Impacts                                           261
    Electric Utilities                                               264
      Structure and Stakeholders                                     264
      Energy Transmission                                            265
      Load Leveling                                                  267
      Load-Leveling Resource Limitations                             270
      Production of Hydrogen for Sale                                 270
    Combined Electric and Gas Utilities                              273
    Nuclear Power—A Key Issue                                        274
    Appendix: Federal Power Commission Priorities
      for Natural Gas Curtailments                                 279
    References                                                       280
    Bibliography on Nuclear Power: Pro and Con                       282

17    STEELMAKING AND AMMONIA SYNTHESIS                             286

    Introduction                                                     286
    Steelmaking                                                      286
      Use of Coal                                                    286
      Steelmaking Processes                                          287
      Impacts                                                        291
      Steel Summary                                                  293
    Ammonia Synthesis                                                294
    References                                                       297

INDEX                                  299

ABOUT THE AUTHORS                       307

LIST OF TABLES

Table                                                                    Page

4.1    Atomic Hydrogen/Carbon Ratio for Some Abundant
       Hydrocarbons                                                       36

4.2    Chemistry of the Steam Reforming of Methane                        36

4.3    Common Gas Nomenclature                                            38

4.4    Westinghouse Hybrid Cycle                                          45

4.5    Comparison of Hydrogen Production Alternatives                     46

5.1    Storage Efficiency of Liquid Hydrogen Containers                   55

5.2    Relative Hydrogen Density of Various Substances                    57

7.1    Comparison of $LH_2$ and Jet–A Passenger Aircraft                  90

7.2    Normal Boiling Points of Cryogenic Liquids                        100

8.1    Properties of Hydrogen and Other Common Fuels
       Relevant to Safety                                                107

9.1    Superconducting Electric Transmission Line Pro-
       jected Materials Needs and Reserve Estimates                      126

10.1   Compatibility of Automotive Technology Options
       with Energy or Fuel Distribution Networks                         133

11.1   U.S. Coal Reserves and Recent Production                          144

11.2   Cost of Producing Hydrogen from Coal                              145

11.3   Production Costs of Electrolytic Hydrogen Excluding
       Electricity Costs                                                 148

11.4   Cost Estimates for Producing Hydrogen from Coal                   150

11.5   Adjusted Cost Estimates for Producing Hydrogen
       by Electrolysis of Water                                          152

11.6   Adjusted Cost Estimates for Producing Hydrogen by
       Thermochemical Decomposition of Water                            153

| Table | | Page |
|-------|---|------|
| 11.7 | Cost of Gaseous Hydrogen Transmission in Optimized Pipelines for Various Assumed Costs of Hydrogen Used to Fuel Compressors | 154 |
| 11.8 | Comparative Investment Costs of Gaseous Hydrogen Storage | 157 |
| 11.9 | Cost Comparisons for Representative Hydrogen Systems | 158 |
| 11.10 | Comparative Costs of Petroleum Products, 1973-74 | 159 |
| 11.11 | Mine Mouth SNG Costs | 160 |
| 12.1 | Natural Building Blocks of a Hydrogen Economy | 167 |
| 14.1 | Hydrogen Generation Needed to Supply 1973 California Automobile Demand | 192 |
| 14.2 | Comparison of Emissions of Air Pollutants in Gasoline-Powered Cars, Petroleum Refineries, and Hydrogen Production Plants | 193 |
| 14.3 | Estimated Investment in U.S. Gasoline Distribution System December 1973 | 202 |
| 14.4 | Societal Impacts of a Transition to Hydrogen-Fueled Private Automobiles and Fleet Vehicles | 207 |
| 15.1 | Net Worth of Airframe Manufacturers, 1972 | 218 |
| 15.2 | Domestic Air Travel by Scheduled Air Carriers: Statistics | 223 |
| 15.3 | Airport Liquid Hydrogen System | 230 |
| 15.4 | Role of Commercial and Civilian Aircraft in U.S. Foreign Trade | 236 |
| 15.5 | Air Pollutant Emissions Comparison: $LH_2$ Versus Jet-A Subsonic Passenger Aircraft | 238 |

Table                                                                      Page

15.6    Noise Comparison:  LH$_2$ Versus Jet-A Subsonic
        Passenger Aircraft                                                  240

15.7    Death Rates from Transportation Accidents                          245

15.8    Summary of Impacts of Hydrogen-Fueled Aircraft                     248

16.1    U.S. Gas Consumption by Region and Class of Ser-
        vice, 1973                                                         259

16.2    Estimated World Reserves of Titanium, 1970                         271

16.3    Estimated Number of Nuclear/Electrolysis Plants
        Needed for Select Sectors                                          275

16.4    Capital Cost Estimate for Nuclear Plant                            276

16.5    Summary of Impacts of Hydrogen Use in Utilities                    278

17.1    Leading Regions or States Producing Iron Ore,
        Bituminous Coal, and Raw Steel, 1972                               288

17.2    Uses of Coal in the Steel Industry, 1970                           289

17.3    Summary of Changes Expected for Direct Reduction/
        Electric Furnace Steelmaking                                       293

17.4    Summary of Impacts of Nuclear/Direct Reduction/
        Electric Furnace Steelmaking                                       295

17.5    Estimated Feedstock Prices That Would Produce
        Hydrogen at Equal Costs (About \$1.50 per 10$^3$
        SCF) by Various Processes                                          296

# LIST OF FIGURES

Figure                                                                    Page

4.1    Typical High-Temperature, Atmospheric-Pressure
       Coal Gasification Process                                           39

4.2    Ideal Electrolysis Cell and Conventional Definition
       of Electrolysis Efficiency                                          41

4.3    Schematic of Closed-Cycle Thermochemical Water
       Splitting and Equations of the Euratom Mark IX
       Cycle                                                               44

5.1    Interstitial Sites for Hydrogen in a Crystal Lattice               58

5.2    Pressure-Composition Isotherms for the $Mg_2Ni$-H
       Metal Hydride System                                                59

5.3    Pressure-Composition Isotherms for the FeTi-H
       Metal Hydride System                                               60

6.1    Natural Gas Transmission Pipelines in the United
       States                                                             67

7.1    U.S. Natural Gas Consumption                                       74

7.2    U.S. Natural Gas Reserves                                          75

7.3    Typical Load Curves for Electric Power Generation
       for One Week                                                       79

7.4    Average Winter Weekly Electric Power Load Curve                    80

7.5    Hydrogen Energy Storage Options for Electric
       Utilities                                                          81

7.6    Options for Using Hydrogen for Energy Transmission
       from Distant Nuclear Reactors                                      83

7.7    Automotive Synthetic Fuels Systems Comparison                      87

7.8    Options for Hydrogen Storage in Aircraft                           92

7.9    Size Comparison: $LH_2$ Versus Jet-A Passenger
       Aircraft                                                           93

| Figure | | Page |
|---|---|---|
| 8.1 | Maximum Vertical Cross-Sections of Flames Produced at Various Time Intervals Following Spillage of 89 Liters of Liquid Hydrogen on a Gravel Surface | 109 |
| 11.1 | Comparison of Price Indexes | 142 |
| 11.2 | Cost of Hydrogen from Coal as a Function of Coal Cost | 146 |
| 11.3 | Estimated Production Costs for Electrolytic Hydrogen as a Function of the Cost of Electricity | 147 |
| 11.4 | Liquefaction Costs as a Function of Electricity Cost | 156 |
| 11.5 | Synthetic Fuels Price Sensitivity to Coal Feed Costs | 161 |
| 12.1 | Conceivable Hydrogen Economy Systems | 168 |
| 13.1 | Optimistic Implementation Scenario | 172 |
| 13.2 | Realistic Implementation Scenario | 173 |
| 13.3 | Strong Government Intervention Implementation Scenario | 174 |
| 13.4 | Comparison of Scenarios for Selected End-Use Sectors | 176 |
| 13.5 | Eras and Events in the Transition to a Hydrogen Economy | 177 |
| 14.1 | Nitrogen Oxide ($NO_x$) Emission from a Small Engine as a Function of Fuel Type | 190 |
| 14.2 | Nitrogen Oxide ($NO_x$) Emission from a Hydrogen-Fueled Automobile Engine Compared to Same Engine Fueled with Gasoline | 191 |
| 14.3 | Energy Resource Utilization of Hydrogen-Fueled and Electric Cars | 204 |
| 15.1 | Major Air Traffic Hubs, 1973 | 215 |

Figure                                                                    Page

15.2     U.S. Domestic Trunks—Scheduled Service, 48-
         State Basis                                                        216

15.3     Progress in Takeoff Noise Reduction                               224

15.4     NEF Contours for a Representative Single-Runway
         Airport in 1970                                                   226

15.5     Resource Utilization Comparison for Synthetic
         Jet Fuel (Kerosene) from Coal and Liquid Hydrogen
         from Coal                                                         242

16.1     Stakeholders in the Gas Industry                                  256

16.2     Industrial Use of Natural Gas, 1968                               261

16.3     Cost of Hydrogen Generation and Transmission as
         a Function of Distance                                            266

16.4     Relative Costs of Transmitting Hydrogen and
         Electricity                                                       267

17.1     Comparison of Conventional Steelmaking and Direct
         Reduction Approach                                                290

# LIST OF ABBREVIATIONS

| | |
|---|---|
| AIF | Atomic Industrial Forum |
| AISI | American Iron and Steel Institute |
| CAB | Civil Aeronautics Board |
| CEQ | Council on Environmental Quality |
| DoD | Department of Defense |
| DoI | Department of the Interior |
| DoT | Department of Transportation |
| EPA | Environmental Protection Agency |
| ERDA | Energy Research and Development |
| FAA | Federal Aviation Agency |
| FAR | Federal Aviation Regulations |
| FPC | Federal Power Commission |
| HEW | Department of Health, Education and Welfare |
| HTGR | High Temperature Gas-Cooled Reactor (Nuclear) |
| IGT | Institute of Gas Technology |
| IRS | Internal Revenue Service |
| $LH_2$ | Liquid Hydrogen |
| LNG | Liquid Natural Gas |
| NASA | National Aeronautics and Space Administration |
| NEF | Noise Exposure Forecast |
| NSF | National Science Foundation |
| NTSB | National Transportation Safety Board |
| OPEC | Organization of Petroleum Exporting Countries |
| OTA | Office of Technology Assessment |
| RANN | Research Applied to National Needs, a Program of the National Science Foundation |
| SCF | Standard Cubic Foot |
| SFO | San Francisco International Airport |
| SMSA | Standard Metropolitan Statistical Area |
| SNG | Substitute Natural Gas |
| SST | Supersonic Transport |

## NOTE TO THE READER

For readers already familiar with the concept of the hydrogen economy who may wish to consult only specific sections of the book, we have tried to ensure that each chapter is as self-contained as possible. Where this would require excessive repetition, we have provided cross-references to other chapters.

Each chapter has a self-contained reference list, so that a number of references are repeated from chapter to chapter. Rather than continually repeat these references in the usual note style, we have simply inserted the number of the relevant reference in brackets within the text. The reader need merely turn to the reference list at the end of the chapter to find the source.

The table of contents and lists of figures and tables are detailed and are intended to assist readers in finding subject matter discussed in several places. For example, the consequences of using hydrogen in aviation described in Chapter 16 should not be read without also consulting the appropriate section of Chapter 7; discussion of hydrogen safety in a particular application should be read in conjunction with Chapter 8.

# MAJOR CONCLUSIONS AND RECOMMENDATIONS

The hydrogen economy concept is broad both in scope and in societal consequences. However, because large changes are involved, the concept could be implemented only slowly even once it were deemed desirable. Accordingly, society has ample time to develop more information and to weigh the advantages and disadvantages of the hydrogen economy concept. The luxury of this long lead time should not be squandered, however, for some of the fundamental course-setting decisions must be made soon.

This section briefly summarizes the major conclusions of the report and presents some general public policy implications. In addition, a list of priorities is given for research and development activities needed to ensure that the knowledge necessary to evaluate future policy decisions becomes available.

## USE OF HYDROGEN

Hydrogen is not a primary energy resource. Instead, it must be regarded as an energy form, or an energy carrier; for like electricity, some other primary energy resource is needed to produce it. Hydrogen can be produced from water and hydrocarbon fuels by thermochemical or electrochemical processes. Today, the demand for hydrogen is low, and hydrocarbons are its major source. However, it is the thrust of the hydrogen economy concept that, in the future, when hydrocarbons are not so readily available, hydrogen would be obtained from water by various means and would replace petroleum products. Again, like electricity, hydrogen could be derived from any primary energy resource, including nuclear, solar, or geothermal resources.

Today, much of the world energy requirement is derived from petroleum or natural gas, two fuels that are highly favored because they are easily transported in liquid or gaseous form. For mobile applications, petroleum products are used almost exclusively because their high energy content per unit weight and volume and their ease of containment greatly facilitate automotive and air travel. Hydrogen, in either gaseous or liquid form, could replace petroleum products or natural gas in essentially every application in which the latter are now used. However, liquid hydrogen is incompatible with existing liquid fuel distribution systems, and there remain some unresolved questions concerning the use of the existing natural gas delivery system for hydrogen.

While some envision an energy economy in the distant future that is essentially all electric (aviation being the most important exception), others envision a hydrogen/electric future in which hydrogen would serve as a common denominator fuel for all mobile and stationary applications in which it proved superior to electricity. Moreover, there would be a cross-link between hydrogen and electricity by virtue of the electrochemical decomposition (electrolysis) of water to make hydrogen and the electrochemical oxidation of hydrogen (in a fuel cell) to make electricity and water.

This examination of the hydrogen economy concept has produced the following conclusions:

   1.  The hydrogen/electric concept of the future energy economy
is technically feasible because hydrogen could replace petroleum or
natural gas in essentially every application.

   2.  The distinctive physical properties of hydrogen (liquid only
at cryogenic temperatures; low energy density; light, bouyant, easily
ignited gas) make its use less convenient than today's commonplace
fuels.

   3.  The use of hydrogen would rarely prove the most economical
alternative in either monetary or basic energy resource terms for the
rest of this century.

   4.  Massive investments in the production, distribution, and
storage of today's fuels pose a very significant barrier to a rapid
voluntary conversion to a hydrogen economy.

   5.  In aviation, hydrogen is probably the superior alternative
fuel.

   6.  Hydrogen would offer many significant environmental benefits at the point of use, but, like electricity, there would still be environmental damage at the point of production.

   7.  Because of the long lead times required, even a vigorous
program to promote the use of hydrogen derived from coal or from
nuclear power and water could not make a significant contribution to
U.S. energy independence before the year 2000.

   8.  The transition from the present fossil-fuel energy economy
to a hydrogen economy would be long—probably a century—because
many existing energy facilities would have to be supplanted.

   9.  A transition to a hydrogen economy could be considered
permanent.  Consequently, the concept deserves consideration commensurate with that given to more temporary energy economy solutions such as synthetic methane or crude oil derived from coal and
oil shale.


## TRANSITION TO HYDROGEN


   Decisions in both the private and public sector are generally
made with a planning horizon of five to ten years.  In the private sector especially, this time horizon derives from the realization that
money has a value that varies with time (even in the absence of inflation).  Corporations formally discount the value of future earnings
anticipated from an investment according to the profit or interest
the same investment could earn elsewhere.  As a result, a dollar
that could be earned this year is considered more valuable than a dollar that could be earned next year.  Discounting of the future tends
to justify a preoccupation with short- rather than long-term goals in
decision making.  During periods when capital formation falls far

short of the quantity desired for new investment, monetary interest rates increase. This in turn steepens the rate at which the long-term future is discounted. In such times, the short term becomes the paramount concern.

Because massive investments are required and technologies with long lead times would have to be deployed, a transition to hydrogen on a large scale would take many decades (probably a century) before it was reasonably complete. As a result, corporations (and governments as well) tend to dismiss hydrogen in favor of those activities that continue the viability of the existing order. Thus, for example, instead of hydrogen, the natural gas industry is concentrating on making synthetic methane from coal, and the petroleum industry is concentrating on making synthetic crude oil from oil shale and coal. This concentration derives from the fact that those materials can be blended with traditional supplies without much perturbation of either the distribution or end-use aspects of the energy system. Although coal gasification and liquefaction and oil shale development require large amounts of capital investment, they offer the advantage of being carried out incrementally without the fundamental alteration of the entire fuel supply system.

Because the transition to hydrogen is genuinely only a long-term option and would take more time to implement than the private sector is normally concerned about, the role of hydrogen in the future U.S. energy economy is rightfully a matter of public policy. The potential effects of the range of governmental attitudes toward hydrogen—neutral, pro-hydrogen, and anti-hydrogen—are considered below:

### Neutral Government Policy

If the federal government remains neutral toward the future of the hydrogen economy, largely through a failure to take a position one way or the other, then the decision about what energy form will ultimately occupy the niche now filled by hydrocarbon fuels will be left to the market mechanism and institutional compatibility as influenced by the results of piecemeal private-sector and federal research and development. In particular:

Hydrogen would not be likely to gain favor until fuel options more compatible with existing corporate, societal, and governmental institutions have been exploited and, perhaps, nearly exhausted. This would make the transition period needlessly rushed and, hence, disruptive.

Hydrogen could not contribute much to the U.S. energy mix before the year 2000, and even then only a few sectors would be affected.

The changeover period would be so long that neither would societal consequences be felt acutely, nor would environmental benefits be realized quickly.

Institutional reluctance to embark on change is likely to impede private-sector research and development spending for a hydrogen economy; as a result, federal research and development spending is needed, if only to ensure that an adequate knowledge base develops to preserve the option of a large-scale transition to hydrogen.

The transition would be paced by necessary improvements in the technology of hydrogen production and their economic consequences compared with the costs of alternative fuels.

The first applications would be those in which hydrogen could be produced and used captively without involving its sale in the energy marketplace.

Applications that depend on the deployment of a new fuel network that would have to compete with an existing and widespread fuel distribution network—such as gasoline for private automobiles and methane gas to residences—would be the last to be implemented.

## Pro-Hydrogen Government Policy

If the federal government were to adopt a pro-hydrogen policy, the total duration of the changeover period might be reduced, but the initial changes could not be speeded up appreciably because of the long lead times involved. The following applications are probably the most readily and quickly affected by a pro-hydrogen policy:

Long-distance domestic commercial aviation (this sector probably could be converted to hydrogen by the year 2000 largely because there already is a high degree of government regulation and participation)

Electric utility load-leveling facilities using hydrogen generation and storage

Nuclear steelmaking with hydrogen used for iron ore reduction

These sectors are the most susceptible because they either involve only hydrogen produced and used captively or require a distribution network with only a few key dispensing locations.

### Anti-Hydrogen Government Policy

If the federal government were to adopt an anti-hydrogen policy, largely manifested by failure to support significant hydrogen research and development efforts while giving support to research and development of other synthetic fuels, the emergence of the hydrogen economy in the United States would probably be delayed until fossil-fuel reserves of all kinds become truly scarce globally. Countries with fewer domestic energy resources than the United States, however, would probably take the research and development lead and begin actual implementation of some aspects of the hydrogen economy concept before the year 2000. The United States could then find itself a buyer of foreign hydrogen economy technology. Even without U.S. governmental research and development support, however, the following components of a hydrogen economy concept would probably begin to be deployed in the United States within 30 years:

Electric utility load leveling using hydrogen generation and storage

Nuclear steelmaking with hydrogen used for chemical reduction of iron ore

The emergence of a hydrogen economy would, of course, be related to other future events. Some of the key events that would play critical roles and therefore should be regarded as signposts are the following:

1. Federal government policy concerning U.S. energy independence. If dependence on foreign sources of oil becomes intolerable and the federal government makes a sustained effort to develop domestic energy resources on a massive scale, the production of hydrogen for mobile applications would probably be stimulated late in the century.

2. Clarification and stabilization of the environmental and occupational health policies of coal mining. In the short term (assuming no additional constraints on nuclear power), coal would be the most economical large-scale primary energy resource for the production of hydrogen.

3. Resolution of the controversies about price regulation of natural gas and imports of liquid natural gas (LNG). Natural gas is one of the key fuels with which hydrogen would have to compete as transition began. If natural gas is deregulated, it is commonly expected that demand will fall while supplies and reserves will rise (owing to increased incentives for exploration). This would both

diminish the size of the gaseous fuel market that hydrogen could ultimately serve and postpone gas industry interest in hydrogen. Im-
portation of LNG could greatly extend the viable lifetime of the natural
gas industry in its present form.

4. Resolution or quieting of the controversies and constraints
on power from nuclear fission (fuel availability after about 1985, long-
term disposal of nuclear wastes, and power plant siting). In the long
term, nuclear power would be one of the major primary energy
sources for hydrogen production. Although coal would be a primary
source in the beginning, should the role of nuclear power be diminished,
then the emergence of a hydrogen economy would probably have to
await economically viable solar energy technologies. In particular,
if nuclear power does not develop as planned, electric power generation would have to become even more dependent on coal than is presently envisioned, and this would tend to limit the coal available for
hydrogen production, thereby impeding the earliest steps toward hydrogen.

In addition to the factors listed above, numerous related technological developments would tend either to enhance or to diminish
the prospects of a transition to hydrogen. These are summarized below:

### Enhance

Economical solar energy collection. Many solar energy
technologies require means to store energy before the energy becomes very useful; storage of solar energy as hydrogen would stimulate delivery of energy in hydrogen form.

Advances in chemical catalysis using abundant low-cost
materials. This would lower the cost and remove the constraint of catalyst availability on electrolyzers and fuel
cells.

Development of practical superconductors with transition
temperatures around 26°K. This would allow use of liquid
hydrogen as a refrigerant and would stimulate advancement
of liquid hydrogen technology.

### Diminish

Technical and economic success of nuclear fusion to
produce electricity. This would tend to advance the evolution of an all-electric economy, thereby diminishing the
role of hydrogen. (However, some feel that it could also
stimulate production of hydrogen with off-peak power.)

Development of long-lived, low-cost, high energy- and power-density electric storage batteries suitable for electric utilities and electric cars. This would diminish the long-term need for hydrogen in mobile applications. In addition, nighttime battery charging would help level electric utility loads, thereby undercutting a major potential captive use of hydrogen that would otherwise be expected to advance the hydrogen technology state of the art.

In general, evolution of a hydrogen energy economy would be more likely to occur in an era of long-term economic prosperity, stable energy and environmental policy, and international peace than in an era marked by depression and uncertainties in energy policy and in the geopolitical sphere. Since unstable energy policy increases the risk inherent in energy technology investment, a lack of stability would strengthen the normal alliance with the "tried and true." Prolonged international instability would probably have a similar effect.

## THE FUTURE OF HYDROGEN

It has been said that nations and individuals inevitably do the right thing, but only after they have exhausted all the other possibilities.* Viewed in the abstract, ultimate conversion to a hydrogen/ electric economy may very well be the long-term "right" thing for the United States. However, the hydrogen economy concept cannot be viewed solely in the abstract. If the concept is ever to be implemented, a feasible pathway must be found linking the present—with its various established institutions, investments, interests, biases, and preferences—to the future. It is not enough that hydrogen might be the best final solution; it would also have to be found the best transitional solution. An institutionally feasible pathway would be characterized by a series of small incremental changes (each of which may, nevertheless, be painful to some established interests) rather than a series of drastic changes.

There is little doubt that as convenient fossil energy resources become exhausted something like the hydrogen economy must eventually take form to accommodate mobile uses of energy. However, there is reason to debate when the transition should or must begin. Should it be decided that hydrogen is a goal to be pursued, then,

---

*Observed, without recalling the source, by Sir Siegmund Warburg in an interview: "Warburg: A European Who Prefers Wall Street," Business Week, November 23, 1974, p. 92.

ideally, interim changes in the energy economy should be in directions that would make a later transition to hydrogen less complex. However, there is a strong possibility that, with a neutral governmental policy toward hydrogen, the series of incremental individual and corporate decisions will take the energy economy in a direction that will make a transition even more difficult than it would be today.*

If research and development are not adequate to remove the uncertainties about hydrogen use and to improve the viability of the hydrogen economy concept, society may well find that it has painted itself into a corner because it had continued to make massive investments in expedient short-term solutions while neglecting to set, and make progress toward, long-term goals.

It is important that a transition to hydrogen as a means of delivering energy to end-users could be permanent, because any primary energy resource developed in the future could be used to produce hydrogen. Thus, in the future, a sequence of basic energy resources could be utilized without ever again affecting consumers.† In this respect, hydrogen offers a clear advantage over synthetic gasoline, diesel, methane, and so on, because ultimately the concentrated carbonaceous resources on which these synthetics are based will be in short supply. Thus, while systems based on synthetic fuels from fossil resources would be easier to implement in the structural framework set by today's culture, they can only be a temporary solution, and yet another transition will be required later.

## RECOMMENDATIONS FOR RESEARCH AND DEVELOPMENT

To preserve the option of choosing hydrogen as an important component of the energy economy of the future, significant research

---

*For example, the development and deployment of coal liquefaction facilities to produce a synthetic crude oil to maintain the viability of the existing petroleum-based system will not make a later transition to hydrogen less difficult. Instead, this only adds to the complexity and to the entrenched investment of the hydrocarbon system. By contrast, investment in solar energy technologies, which will require means of energy storage (to tide over nighttime and unfavorable insulation conditions), would tend to advance the use of hydrogen storage systems and would also spur the delivery of solar energy in hydrogen form. Therefore, this investment would tend to make a future transition to hydrogen easier.

†Similarly, electric generation will soon increase the diversity of its primary energy sources by using solar and geothermal energy without affecting consumers.

and development should be undertaken. However, the priorities for research and development options should take into account the transitional sequencing of the introduction of hydrogen production and utilization technologies that reflect real-world constraints in the existing institutional infrastructure. As a result, some potential research topics, such as further characterization of air pollutant emissions of internal combustion engines, could be shelved while awaiting progress in other, more fundamental areas such as low-cost, high-efficiency hydrogen production.

If the lesson of the initiation and subsequent cancellation of the SST program teaches anything, it is that a clear distinction can and should be made between the decision to "explore" and the decision to "deploy" a technological option. The following table gives a list of research and development topics (broken into three priority groups) that reflects both the need to explore some topics and the estimated criticality of any given topic in later decisions to deploy aspects of a hydrogen economy. The agency of government or the sector of private industry that seems best suited for leadership is indicated; the emphasis and priorities are taken to reflect the federal government point of view.

| Priority | Sponsor |
| --- | --- |
| High (to begin at once) | NSF |
| 1. Development of advanced concept hydrogen production technologies, especially high-pressure, high-temperature electrolysis and closed-cycle thermochemical processes. This is the most critical area, for without considerable cost reductions, all other questions about hydrogen are moot. | ERDA Industry |
| 2. Development of advanced materials suitable for containing the high-temperature, corrosive chemicals to be used in closed-cycle thermochemical processes. Without concurrent work on this topic, progress on part of topic 1 would be difficult. | NSF ERDA |
| 3. Investigation of hydrogen-environment embrittlement in materials expected to be used in the hydrogen economy under realistic conditions of temperature, pressure, and hydrogen purity. The results of such an investigation are essential to avoid the study of inappropriate concepts and to contribute to understanding of real hydrogen safety. | NSF ERDA NASA DoD |

| Priority | Sponsor |
| --- | --- |
| 4. Rigorous analytical and experimental evaluation of hydrogen safety in likely applications in realistic environments under realistic operational conditions. Once success was indicated in item 1, safety would probably become the number one public concern. | ERDA DoT HEW DoD |
| 5. Determination (by social scientists) of the baseline public attitude toward the use and safety of hydrogen, followed by periodic interviews (at five- to ten-year intervals) to ascertain how and why attitudes toward hydrogen change. This research should begin at once, because valuable baseline data may already be lost now that the movie Hindenburg is released. This information would be needed to develop a public education program to disseminate the results of topic 4. | NSF-RANN HEW |
| 6. Development of the air and ground systems needed to support hydrogen-fueled commercial passenger aviation. Research on this topic is apparently about to begin with NASA funding; it should continue. This is a high-priority topic because aviation is one of the best candidates for early transition to hydrogen. | NASA |
| 7. Systems studies and detailed technology assessment of the introduction of hydrogen into commercial aviation. This study should include preparation of detailed implementation scenarios with the assistance of stakeholders. Because massive government involvement seems essential and the lead times are long, key decisions may be needed in the next six to ten years if this sector is to use hydrogen in the 1990s. | NASA DoT NSF-RANN |
| 8. Evaluation of possible temporary or absolute materials resource limitations of the hydrogen economy. The results of these studies are needed to guide development of advanced processes for hydrogen production and consumption. This is a key topic intended to ensure that research and development money is spent only on technologies that could actually be deployed on a large scale. | DoI ERDA NSF-RANN |
| 9. Systems modeling of the U.S. energy economy to produce scenarios of hydrogen introduction that takes into account hydrogen cost, interfuel competition price relationships, environmental pro- | ERDA NSF-RANN EPA CEQ |

| Priority | Sponsor |
| --- | --- |

tection, and institutional constraints. No completely adequate model yet exists to guide policy
making concerning the appropriate sequencing of
conversion of energy use sectors to hydrogen.
Results of this effort would allow development of
a clear set of hydrogen research and development
priorities.

10. Technological development of hydrogen reduction — Industry
of iron ore. Although this topic could be expected
to contribute to the advancement of hydrogen production, environmental cleanup is the primary
benefit.

Medium (to begin at once when encouraging results
have been obtained from high-priority topics)

1. Examination of approaches to convert local — Industry
mothane distribution pipelines to hydrogen — FPC
use. — ERDA

2. Development of hydrogen energy storage sys- — Industry
tems suitable for use by electric utilities for — ERDA
load-leveling applications. This is proceeding
already with some industry funding.

3. Further examination of the institutional and — OTA
economic barriers to implementation; descrip- — ERDA
tion of the incentive options that could surmount — NSF
the barriers. This would amount to an in-depth
technology assessment of the concept.

4. Optimization studies of hydrogen economy build- — ERDA
ing blocks based on the new research and devel- — NASA
opment results obtained in other topics mentioned. These would provide important input to
the in-depth technology assessment suggested
immediately above.

5. Demonstration flights of a cargo airplane con- — NASA
verted to liquid hydrogen fuel to gather operating experience and to test public acceptance of
hydrogen.

Low (to be held in abeyance until actual deployment
is closer and the configuration of hydrogen technologies is better defined)

1. Further demonstrations that existing automobile — EPA
engines can be operated on hydrogen and mea- — ERDA
surement of their pollutant emissions.

| Priority | Sponsor |
|---|---|
| 2. Study of hydrogen systems that use materials too scarce ever to be useful on a large scale (for example, noble-metal catalysts, rare-earth metal hydrides) unless these systems offer a advantage as a testing ground for studying general concepts or for obtaining fundamental scientific understanding. | All sponsors |

# TECHNOLOGY ASSESSMENT AND
# ENERGY IN THE FUTURE

Technology assessment as a public endeavor is still young but is gaining attention. Now that several assessments have been attempted by various research teams, it is recognized that there can be no single appropriate method to approach such broad-ranging studies. Rather, each technology (or family of technologies) requires a special tailoring of the study techniques to be applied. Only in this way can this form of study be responsive to the wide variation in the breadth and depth of societal consequences resulting from diverse technologies.

This section briefly discusses the origins of technology assessment, some generalizations about the future of energy in this country, and the methods applied in this study.

# 1

## THE CONCEPT OF
## TECHNOLOGY ASSESSMENT

The idea that many of the broad consequences of technological change might be anticipated in advance of their actual occurrence has gained acceptance in the past several years. Broad systematic anticipation of future social, environmental, institutional, and economic effects is a relatively new endeavor, although forecasting in narrow subject or business areas has long been practiced. The process of attempting to foresee such impacts is now called technology assessment [1-9].* The first true technology assessments were sponsored and commissioned by the Research Applied to National Needs (RANN) Program of the National Science Foundation [10, 11]. In 1974, the Office of Technology Assessment (OTA), a new government organization reporting to the U.S. Congress, also began to commission technology assessments [10, 12, 13].

Generally, the goals of a technology assessment are to:

1. Improve the level of information available to decision makers about the direct and indirect consequences of technological change

2. Provide a point of departure for parties having a stake in the outcome of the change to voice their perceptions, attitudes, and concerns

3. Identify policy or decision options that could improve the outcome of technological change by maximizing the beneficial and minimizing the detrimental aspects

---

*The process would be more generally understood if instead the name were technology impact assessment.

These goals have been articulated many times in similar statements by many people sponsoring or endeavoring to perform a technology assessment.

The audience for technology assessments is broad and steadily increasing. It is generally recognized that the audience includes decision makers at various levels of government (legislative and administrative); decision makers in the private sector; and the many publics, or groups, in the population with a stake in the consequences. In addition, a long-overlooked, but important, audience is the technologists actively involved in development who frequently have not had the opportunity or inclination to assess the consequences of their professional activities. The recent creation of OTA demonstrates that Congress has become aware of the increasing need to gauge the possible ramifications of legislative actions that increasingly concern technology [2].

The time scale for the evolution of technological change varies greatly depending upon the technologies involved. Some technologies possess the capability to cause great change in just a few decades (for example, jet-powered commercial aviation, transistors, and birth control pills). Other technologies have very long developmental lead times or protracted deployment schedules, and their impacts develop slowly or become important only long after the technology is originally conceived (for example, electric power, video telephones, and space travel). The concept of an economy that largely relies upon the use of hydrogen produced from primary energy sources to deliver energy falls into the latter category. An economy fully dependent on hydrogen* is not likely ever to be reached, and widespread hydrogen use probably cannot reach its full flower before the middle of the twenty-first century.

The performance of a technology assessment for such a distant technology is handicapped both by the major uncertainties about the general state of society that far in the future and by uncertainties about the form of the technology that will actually be deployed. As a result, for distant technologies descriptions of societal impact must necessarily be rather general and imprecise, and the public policy considerations must also remain rather general. For the hydrogen energy economy, the presently relevant policy issues are generally related to research and development activities that could reduce uncertainties and keep open the option of an eventual transition to a hydrogen energy economy. Accordingly, although this study is a preliminary technology assessment, such an early assessment provides an improved chance to plan, to adapt, and to steer the evolution of the technology in question.

---

*Or any other single energy form, such as electricity.

## REFERENCES

1. J. F. Coates, "Technology Assessment: The Benefits, the Costs, the Consequences," The Futurist, December 1971, pp. 225-31.

2. Technology: Process of Assessment and Choice, report of the National Academy of Sciences to the Committee on Science and Astronautics, House of Representatives, 91st Cong., 1st sess., July 1969.

3. A Study of Technology Assessment, report of the Committee on Public Engineering Policy, National Academy of Engineering, to the Committee on Science and Astronautics, House of Representatives, 91st Cong., 1st sess., July 1969.

4. Technical Information for Congress, report by the Science Policy Research Division, Legislative Reference Service, Library of Congress, to the Subcommittee on Science, Research and Development of the Committee on Science and Astronautics, House of Representatives, 91st Cong., 1st sess., 1969.

5. R. G. Kasper, ed., Technology Assessment: Understanding the Social Consequences of Technological Applications (New York: Praeger, 1972).

6. H. Folk, "The Role of Technology Assessment in Public Policy," in Technology and Man's Future, ed. A. H. Teich (New York: St. Martin's, 1972).

7. H. Brooks and R. Bowers, "Technology: Process of Assessment and Choice," in Technology and Man's Future, ed. A. H. Teich (New York: St. Martin's, 1972).

8. H. Brooks and R. Bowers, "The Assessment of Technology," Scientific American, February 1970, pp. 13-21.

9. M. V. Jones et al., "Technology Assessment," The Mitre Corporation, Washington, D.C., June 1971.

10. E. M. Dickson and R. Bowers, The Video Telephone: Impact of a New Era in Telecommunications (New York: Praeger, 1974).

11. J. F. Coates, "Technology Assessment and Public Wisdom," Journal of the Washington Academy of Science 65, no. 1 (1975): 3-12.

12. J. F. Burby, "Infant OTA Seeks to Alert Congress to Technological Impacts," National Journal Reports, September 21, 1974, pp. 1418-29.

13. J. F. Burby, "OTA Works to Produce Track Record with Six Major Projects," National Journal Reports, September 28, 1974, pp. 1454-64.

# 2

**ENERGY AND HYDROGEN
IN THE FUTURE**

## CHANGE IS INEVITABLE

Few people would dispute that the world is entering an era of increased competition for the remaining fossil-fuel resources, especially petroleum and natural gas [1]. Higher standards of living and increases in population increase the demand for energy. Moreover, as the richest metallic mineral resources are depleted and ever leaner ores are mined and processed, energy must be consumed at an increasing rate just to sustain materials production for an industrial society. The day when technologies able to exploit unconventional sources of energy must be developed and deployed comes ever closer [1].

There are many options available for meeting or altering future energy demands in both the near and long terms, including the following:

Introduction of energy conservation measures
    Improved conversion efficiencies
    Improved end-use efficiencies
    Reduction of wasteful practices
Extended development of petroleum and natural gas supplies by
    Tertiary recovery
    Discovery and development on outer continental shelves and
      under the deep sea
    Discovery and development in remote land environments (for
      example, Alaska)
Increased reliance on coal of various grades for
    Direct combustion
    Liquefaction into portable fuel
    Gasification into pipeline-quality fuel

Production of synthetic liquid fuels from unconventional fossil hydrocarbons
  Oil shale
  Tar sands
Utilization of carbonaceous wastes
  Municipal and industrial
  Forest or agricultural
  Sewage sludge
Increased use of nuclear fission reactors for electricity generation
  Conventional water-cooled
  High-temperature gas-cooled
  Breeder reactors
Development of nuclear fusion
  Magnetic confinement of plasmas
  Laser-induced
Application of solar energy in many direct and indirect forms
  Sunlight
  Wind
  Falling water (for example, hydroelectric)
  Ocean temperature differences
  Biomass grown for use as a fuel
Generation of electricity from geothermal energy
  Natural steam and hot water reservoirs
  Dry hot rock

Each of these approaches to developing energy has its own local, temporal, environmental, or economic advantages and disadvantages (indeed, the feasibility of some has not been demonstrated). No single approach is adequate or appropriate for all circumstances. Certainly in the future, as today, a multiplicity of resources, technologies, and techniques for providing energy will be used simultaneously.


## DIFFERENCES BETWEEN THE PAST
## AND THE FUTURE

The future will be different from the present in four very important ways. First, the world now depends almost entirely on fossil fuels, especially petroleum, natural gas, and coal, which are readily storable and transportable (although coal, being a solid, is much less convenient for most applications than liquid petroleum products). However, increasingly, primary energy will be produced in forms that are neither portable nor storable. As a result, new emphasis will be placed on technologies that render unwieldy basic energy re-

sources into convenient forms. Two prime examples are the proposed production of gaseous fuels from coal and the planned expansion
in the use of nuclear fission energy to generate electricity. Just as
use of a multiplicity of basic energy resources can be expected in the
future, so a multiplicity of energy storage and carrier techniques can
be expected.

A second important way in which the future will differ from the
present is that the large-scale winning of energy from a basic resource, be it sunlight or underground fossil-fuel deposits, will increasingly occur in places distant from the location of final demand.
Often, as illustrated by Arabian oil, the basic energy resources will
be beyond the political influence or protection of the U.S. government.
This, no doubt, will continue to give rise to sentiment favoring energy
alternatives that would offer independence from foreign control as in
the planning for "Project Independence" [1]. As International Political
situations ebb and flow, so certainly will the sentiment for energy independence.

A third difference between the past and the future, for the United
States at least, is the strong likelihood that continued emphasis will
be placed on environmental quality, especially as it relates to energy
production and end-use. Pollution abatement procedures are often
more practical when undertaken on a large scale at a central facility
than on the small scale that is characteristic of dispersed end-uses.
Therefore, strong pressures will exist to meet the twin needs of energy storage and portability with the energy forms that are cleanest
in their end-use characteristics.

A fourth important difference is the certainty that some concentrated deposits of carbonaceous materials of fossil origin will become
physically exhausted and others will become uneconomical to recover.
When such events begin to fall within planning horizons, strong incentives will arise for the development of nonfossil energy resources and
completely synthetic fuels [1]. Ideal fuel or energy carriers would
be sought that would be derived from inexhaustible, ubiquitous materials; convenient and efficient in production, distribution, storage,
and use; and environmentally clean in combustion or alternative uses.
Because the United States possesses enormous resources and reserves
of coal and oil shale, the time when domestic fossil resources are
physically exhausted is centuries away [1]. However, less fortunately
endowed nations (such as Japan) may be forced to consider nonfossil
synthetic fuels much earlier, especially if depletion of their own resources is coupled with their own desire to achieve a measure of energy independence.

## THE POSSIBLE ROLE OF HYDROGEN

When the criteria for an ideal fuel are considered together in the abstract—inexhaustibility, cleanliness, convenience, independence from foreign control—it often seems to many people as if nature intended mankind to use hydrogen as a fuel. In particular, hydrogen possesses these properties [2-9]:

1. Derivable from any primary energy source
2. Obtainable from water, a common substance
3. Natural precipitation would recycle water between consumption and production
4. Convenient to transmit and consume in gaseous form (the physical state at ambient temperatures and pressures)
5. Possible to transport and store as a cryogenic liquid (very low temperature)
6. Clean on combustion (oxidation), with harmless water the major exhaust product

Of course, there are qualifications to the above statements that make hydrogen, in reality, somewhat less than the ideal fuel. Nevertheless, it is easy to understand why numerous scientists, engineers, environmentalists, and journalists, especially since 1971, have vigorously promoted the concept of the hydrogen energy economy [1-18]. Much of their enthusiasm for the hydrogen economy concept seems to have originated in a search for a solution to automotive air pollution [10-24]. Although "clean air" would be an important attribute of the hydrogen energy economy, other, more fundamental, attributes are even more enticing.

Before discussing the advantages of the hydrogen energy economy concept, it must be emphasized that hydrogen is not a basic energy resource. Hydrogen cannot be found chemically free in nature, and it cannot be obtained without the expenditure of some other energy form. Therefore, hydrogen must be considered an "energy carrier" —a means to transport and store energy derived from other sources. In this respect it would be analogous to electricity [2].

## ADVANTAGES OF HYDROGEN

Probably the most attractive advantage in the use of hydrogen as an energy carrier comes from the ability to manufacture hydrogen from water using any primary energy resource. This means that hydrogen could serve as a chemical common denominator in the energy economy. Electricity plays a similar role because many diverse

primary energy resources can be used to generate it. In fact, the
consumer has no way of knowing, and little reason to care, whether
the electricity he uses was derived from falling water, coal combustion, wind, sunlight, nuclear power, or combustion of garbage. The
common denominator aspect of electricity has greatly simplified matters for the consumer because the burden of any change or adjustment
has been placed on the electric power utilities. Although these utilities do not make use of the full diversity of possible generation
sources today, in the future they will probably be using various forms
of solar energy and geothermal energy not now generally feasible [1].
Currently there is no common denominator in the total fluid or gaseous
fuels economy. It could perhaps be argued, however, that in some
major sectors of the economy a particular fuel essentially acts as a
common denominator (for example, gasoline is the common fuel for
cars of all makes and national origin). The possibility that a single
fuel such as hydrogen could assume a common denominator role in the
portable fuels arena analogous to the use of electricity for stationary
applications is attractive. It is especially so since hydrogen is derivable from water, which is an inexhaustible resource, and, hence,
a transition to hydrogen could be permanent.

An important aspect of using hydrogen as a common denominator
fuel comes from its close relationship with electricity, the other common denominator energy form. As will be described later, hydrogen
is most readily obtained from water by electrolysis; in reverse, hydrogen can be used to generate electricity either by combustion or in
a device called a fuel cell. Thus, if hydrogen were to become a common denominator fuel, there would be a great degree of interchangeability between the two common denominator forms of energy. This
interchangeability would offer many opportunities for economic, technological, and social benefit [3].

## THE NEED FOR A TECHNOLOGY ASSESSMENT

In spite of the apparent attractiveness of hydrogen, its eventual
widespread use is by no means assured. Its characteristics are very
different from the forms of energy to which people are accustomed,
and there are near-term alternatives to it. People will evaluate the
physics, chemistry, economics, convenience, safety, implied institutional change, and so forth of all available alternatives. Hydrogen
may well be viewed as inferior to the alternatives and be rejected.

At an early stage, before either hydrogen or its alternatives
are selected for eventual adoption, a technology assessment would be
useful in illuminating various implications of a transition to a full or
partial hydrogen energy economy. This study is the beginning of
such an assessment effort and seeks to identify:

1. Technical barriers to the production, distribution, and use of hydrogen

2. Benefits and risks

3. Stakeholders (both knowing and unknowing) who have an interest in the forms of energy used in the future

4. Uncertainties that can affect the forms of energy used

5. Critical factors that will dominate decision making

6. Strategies that can reduce uncertainties so that improved evaluations and decisions can be made

A hydrogen energy economy is constrained to a slow evolution because enormous changes face the nation in the area of energy; massive investment has already been made in hardware for conventional energy activities; and huge investment would be needed before significant amounts of hydrogen could be employed. Established infrastructures are resistant to change, and the temptation is continually strong to make only minor incremental modifications that will extend the lifetime of the existing order in preference to undertaking a major change.* Because of perpetual neglect, the state of the art in alternative systems usually lags far behind that of the dominant existing system, sometimes at the price of making even more difficult a transition that is recognized as ultimately necessary. Consequently, this preliminary assessment is largely concerned with the nature and implications of the transition process itself and the actions or policies that could help to preserve the option of an eventual transition to the hydrogen economy.

REFERENCES

1. U.S., Federal Energy Administration, Project Independence (Washington, D.C., November 1974).

2. D. P. Gregory et al., "A Hydrogen-Energy System," American Gas Association, Alexandria, Va., August 1972.

---

*An example of this resistance to change can be found in the dominance of petroleum-derived gasoline in the private automobile sector. To sustain the existing distribution, manufacturing, and end-use investments in the face of declining petroleum availability, the liquefaction of coal and conversion of oil shale to produce a synthetic gasoline is now receiving serious evaluation. However, relatively little consideration is being given to alternatives, such as electric cars powered by electricity derived from the same coal, that do not preserve the gasoline distribution system and consumer investments.

3. "Hydrogen: Likely Fuel of the Future," Chemical and Engineering News, June 26, 1972, pp. 14-17.

4. "Hydrogen Fuel Use Calls for New Source," Chemical and Engineering News, July 3, 1972, pp. 16-18.

5. "Hydrogen Fuel Economy: Wide Ranging Changes," Chemical and Engineering News, July 10, 1972, pp. 27-29.

6. D. P. Gregory, "A New Concept in Energy Transmission," Public Utilities Fortnightly, February 3, 1972, pp. 3-11.

7. D. P. Gregory, "The Hydrogen Economy," Scientific American, January 1973, pp. 13-21.

8. T. H. Maugh II, "Hydrogen: Synthetic Fuel of the Future," Science 178 (November 24, 1972): 849-52.

9. "Hydrogen and Other Synthetic Fuels," a summary of the work of the Synthetic Fuels Panel, prepared for the Federal Council on Science and Technology R&D Goals Study, September 1972.

10. L. Lessing, "The Coming Hydrogen Economy," Fortune, November 1972, pp. 138-46.

11. "Fuel of the Future," Time, September 11, 1972, p. 46.

12. "The Wonder Fuel," Newsweek, November 12, 1973, p. 75.

13. L. W. Jones, "Liquid Hydrogen as a Fuel," Science 174 (October 22, 1972): 367-70.

14. "When Hydrogen Becomes the World's Chief Fuel," Business Week, September 23, 1972, pp. 98-102.

15. W. Clark, "Hydrogen May Emerge as the Master Fuel to Power a Clean Air Future," Smithsonian, August 1972, pp. 13-18.

16. G. D. Brewer, "The Case for Hydrogen-Fueled Transport Aircraft," Astronautics and Aeronautics, May 1974, pp. 40-51.

17. L. T. Blank et al., "A Hydrogen Energy Carrier," Systems Design Institute, National Aeronautics and Space Administration-American Society for Engineering Education, Johnson Space Center, Houston, Texas, 1973.

18. L. W. Jones, "Liquid Hydrogen as a Fuel for Motor Vehicles: A Comparison with Other Systems," Seventh Intersociety Energy Conversion Engineering Conference, 1972, pp. 1364-65.

19. R. G. Murray, R. J. Schoeppel, and C. L. Gray, "The Hydrogen Engine in Perspective," Seventh Intersociety Energy Conversion Engineering Conference, 1972, pp. 1375-81.

20. M. R. Swain and R. R. Adt, "The Hydrogen-Air Fueled Automobile," Seventh Intersociety Energy Conversion Engineering Conference, 1972, pp. 1382-87.

21. W. J. D. Escher, "On the Higher Energy Form of Water ($H_2O^*$) in Automotive Vehicle Advanced Power Systems," Seventh Intersociety Energy Conversion Engineering Conference, 1972, pp. 1392-402.

22. L. O. Williams, "The Cleaning of America," <u>Astronautics and Aeronautics</u> 10, no. 2 (February 1972): 42.

23. P. B. Dieges et al., "An Answer to the Automotive Air Pollution Problem...The Hydrogen and Oxygen Fueling System for Standard Internal Combustion Engines," First Annual Report of the Perris Smogless Automobile Association, Perris, Calif. (undated).

24. W. J. D. Escher, "The Case for the Hydrogen-Oxygen Car," in <u>The Analog Science Fact Reader</u> (New York: St. Martin's, 1974).

# 3

## METHODS OF THE STUDY

The technique used in this study is a semiquantitative form of systems analysis. The construction of simple systems diagrams of the various technologies of a possible hydrogen energy economy and their possible place in U.S. society made it possible to trace the flow of materials, energy, and money through the relevant energy systems. This process greatly aided the identification of stakeholders (both those who know of their future involvement and those who do not), transactions, the things of value transacted, the forces or institutions that regulate them, and the factors on which decisions are likely to be based.

Seven major resources served this study:

1.  The vast body of literature on a hydrogen economy that has developed in the last few years*

2.  Four national meetings at which the hydrogen economy was either a major or the sole topic

3.  Discussions at these meetings with many authors of relevant papers in the literature

4.  Discussions with experts and visits to installations

---

*This literature, however, turns out to be largely secondary in nature. Only a very few primary references seem to have spawned a great flurry of cross citations. The number of papers with something truly new to say has actually been rather small until this past year (1974–75). The bibliography presented at the end of this chapter lists some of the most comprehensive reference materials.

    5.  Services of a consultant* widely known among hydrogen economy enthusiasts who offered critical comments and facilitated personal contacts

    6.  Discussions with professionals on the SRI staff with pertinent expertise

    7.  Simultaneous involvement in a technology impact assessment of deriving synthetic liquid fuels (oils and methanol) from coal and oil shale

It became clear early in the study that a major component of the study would be development of scenarios depicting the possible implementation of the hydrogen energy economy concept.  It was also clear that these scenarios had to be more realistic than the wishful thinking that pervades the literature.  As will be detailed in Chapter 1, after the technical and economic background has been presented, the implementation of even a portion of a hydrogen energy economy would be slow and costly and would involve institutional changes. Advocates of a hydrogen economy commonly fail to distinguish between end-points (when the conversion of specific sectors is essentially completed) and the period of transition.  It was readily determined early in the study that the end-points are a very long time away (well into the twenty-first century).  Consequently, most of the study had to be concerned with transitions in various sectors.  However, description and consideration of the end-points proved to be a useful guide to the form and impact of the transition process.

    Technologies encounter physical limitations and exhibit economies of scale.  The optimally sized unit, or "natural building block," characteristically has reaped nearly all of the economy-of-scale benefits possible and has been standardized in its manufacture.  Rather than build a device or system twice as large as the natural building block, two building blocks are built in parallel.  Good examples of the building-block concept are railroad boxcars, petroleum storage tanks, and 1,000-Megawatt (equals 1 Gigawatt [GW]) nuclear reactors. To formulate the transition scenarios, it proved useful to describe the physical systems in terms of natural building blocks (see Chapter 13).

    Once system components have been described in terms of natural building blocks, it is often found that conceivable systems are not practical.  In particular, a system that seems sensible when considered only as "A feeds into B, which feeds into C" loses credibility once it is observed that the natural sizes of A and C do not match the

---

*W. J. D. Escher of Escher Technology Associates, Inc., St. John's, Michigan.

natural size of B. Mismatches between the natural building-block components of the hydrogen energy economy were found in abundance; this suggests that the implementation of the concept may be greatly impeded. Thus, the use of the natural building blocks in systems diagrams have been used to identify physical bottlenecks, institutional barriers or conflicts, and economic barriers.

In Chapter 13 scenarios will be shown that depict three implementation conditions:

1. <u>Optimistic</u>, assuming no government intervention, ignoring hydrogen's economic disadvantage, and assuming minimum decision and technology lead times

2. <u>Realistic</u>, assuming some government involvement in decision making, some improvement in hydrogen's cost relative to alternatives, and less optimistic lead times

3. <u>Strong government intervention</u>, assuming the federal government mandates the use of hydrogen, backed by appropriate barriers and incentives

These scenarios make clear that within the time frames that can be sensibly foreseen, a hydrogen economy remains in a "transition period." These scenarios form the basis for this study's analysis of societal impacts that have been identified by reflection on the structure of the economy and society as well as by observations of societal response to suggestively similar events.

BIBLIOGRAPHY

Major Information Resources

Blank, L. T., et al. "A Hydrogen Energy Carrier." Systems Design Institute, National Aeronautics and Space Administration-American Society for Engineering Education, Johnson Space Center, Houston, Texas, 1973.

Brewer, G. D., et al. "Study of the Application of Hydrogen Fuel to Long Range Subsonic Transport Aircraft." National Aeronautics and Space Administration, Langley Research Center, Hampton, Va., January 1975.

Fein, E. "A Hydrogen Based Energy Economy." The Futures Group, Glastonbury, Conn., October 1972.

Gregory, D. P., et al. "A Hydrogen-Energy System." American
    Gas Association, Alexandria, Va., August 1972.

"Hydrogen and Other Synthetic Fuels." A summary of the work of the
    Synthetic Fuels Panel, prepared for the Federal Council on
    Science and Technology R&D Goals Study, September 1972.


Major Collections of Papers

Linke, S., ed. Proceedings of the Cornell International Symposium
    and Workshop on the Hydrogen Economy. Ithaca, N.Y.: Cor-
    nell University, April 1975. Forty-five papers.

Ninth Intersociety Energy Conversion Engineering Conference, San
    Francisco, Calif., August 26-30, 1974. Nine papers.

Seventh Intersociety Energy Conversion Engineering Conference, San
    Diego, Calif., September 25-29, 1972. Seven papers.

Veziroglu, T. N., ed. Hydrogen Energy. New York: Plenum, 1975.
    Eighty papers.

_____. Hydrogen Energy Fundamentals: A Symposium Course, Six-
    teen papers presented at the University of Miami, Fla., March
    3-5, 1975.


Semipopular Reviews

Gregory, D. P. "The Hydrogen Economy." Scientific American,
    January 1973, pp. 13-21.

"Hydrogen Fuel Economy: Wide Ranging Changes." Chemical and
    Engineering News, July 10, 1972, pp. 27-29.

"Hydrogen Fuel Use Calls for New Source." Chemical and Engineer-
    ing News, July 3, 1972, pp. 16-18.

"Hydrogen: Likely Fuel of the Future." Chemical and Engineering
    News, July 3, 1972, pp. 16-18.

Lessing, L. "The Coming Hydrogen Economy." Fortune, November
    1972, pp. 138-46.

Maugh, T. H., II. "Hydrogen: Synthetic Fuel of the Future."
    Science 178 (November 24, 1972): 849-52.

Winsche, W. E., et al. "Hydrogen: Its Future Role in the Nation's
    Energy Economy." Science 180 (June 29, 1973): 1325-32.

Abstracts and Listing of Literature

Cox, K. E., ed. Hydrogen Energy: A Bibliography with Abstracts.
    Albuquerque: The Energy Information Center, University of
    New Mexico, January 1, 1974.

Hydrogen, Future Fuel. Boulder, Colo.: Cryogenic Data Center,
    National Bureau of Standards, quarterly.

# HYDROGEN TECHNOLOGIES—
# PRESENT AND PROJECTED

In describing hydrogen economy technologies and conceivable systems, we have tried to strike a balance between technical complexity and simplicity so that this single volume may be informative to those with a wide range of backgrounds. Excellent descriptions of much of the following material can be found in the literature, but we feel it is important that this book be self-contained. Readers who desire or need more technical depth than that presented here should consult the references.

# 4

## HYDROGEN PRODUCTION

### SOURCES OF HYDROGEN

Because hydrogen is chemically very reactive, it is not found in its elemental state on Earth. However, combined chemically with other elements, it is present in water (the most abundant hydrogen resource), fossil hydrocarbons (coal, petroleum, natural gas, and oil shale), biological materials (carbohydrates, protein, and cellulose), and minerals (such as bicarbonate rocks). Energy must be supplied to release hydrogen from any of these compounds by breaking the chemical bonds. Thus, hydrogen is not a primary energy resource obtainable from nature in the same manner as petroleum or coal. Instead, hydrogen is properly regarded as an energy carrier, or a means to store and transmit energy derived from a primary energy resource. Methods of obtaining hydrogen are described in this chapter.

### HYDROGEN PRODUCTION FROM FOSSIL FUELS

Fossil hydrocarbons contain hydrogen and carbon atoms in varying ratios and can be used as a source of hydrogen [1]. Table 4.1 shows the atomic hydrogen to carbon ratio for several of the more important and abundant hydrocarbons [1]. Ironically, the most favorable hydrogen to carbon ratio is found in methane, $CH_4$, the major component of natural gas, a fuel that hydrogen might ultimately supplant. Today, relatively little hydrogen is used compared to the requirements of a mature hydrogen economy, and much of it is obtained by decomposing methane and steam by a process called "reforming" [1]. Half of the hydrogen is derived from the steam.

TABLE 4.1

Atomic Hydrogen/Carbon Ratio for Some Abundant Hydrocarbons

|  | Atomic H/C Ratio |
|---|---|
| Methane ($CH_4$) | 4 |
| Petroleum (heavy fuel oil) | 1.5–1.6 |
| Oil shale | 1.6 |
| Coal | 0.72–0.92 |

Source: Unless otherwise noted, tables and figures are compiled by the authors.

Reformation of methane consists of a series of chemical reactions involving methane, water (steam), and a catalyst; heat must be supplied. In effect, the hydrogen is stripped from both the methane and the water molecules; the reject carbon and oxygen are discarded in the form of carbon dioxide, $CO_2$. Reformation processes can be termed to be "open-cycle" thermochemical production methods. The chemical steps for reforming methane are shown in equation form in Table 4.2. This reforming process is only about 70 percent efficient when all the energy inputs needed to achieve the complete reformation are considered. Other hydrocarbons can also be reformed to produce hydrogen by similar processes.

Hydrogen produced from methane is the source of the liquid hydrogen rocket propellant used by the U.S. space program and gaseous hydrogen used to synthesize ammonia [2]. About 75 percent of ammonia is used in agriculture for fertilizer [2]. Because reserves of methane are declining while demand continues to increase, methane cannot be considered available for long-term, large-scale production of hydrogen.

TABLE 4.2

Chemistry of the Steam Reforming of Methane

| Reforming reaction | $CH_4 + H_2O \rightarrow CO + 3\ H_2$ |
|---|---|
| Shift reaction | $CO + H_2O \rightarrow CO_2 + H_2$ |
| Net reaction | $CH_4 + 2\ H_2O + \text{heat} \rightarrow CO_2 + 4\ H_2$ |

Hydrogen obtained from coal is expected to be the lowest-cost large-scale source of hydrogen (not based on oil or methane) for many years (see Chapter 11). Coal gasification must be considered one of the key technologies if implementation of a hydrogen economy is to begin in this century. Although coal gasification was developed years ago, it is currently undergoing improvement and is about to be deployed in the first modern, large commercial plants. However, most research and development effort and investment commitment to coal gasification have been directed toward the production of methane for use as a substitute natural gas (SNG).* Thus, technical modifications needed to produce hydrogen instead of methane are relatively simple and well understood.

The impetus for SNG development came from the natural gas industry, especially the pipeline companies and utilities, as a means to offset declines in the supply of domestic natural gas relative to demand. The emphasis has been on the production of methane rather than hydrogen because SNG could be mixed into the supplies of natural gas directly. This would result in an essentially unchanged product delivered to the consumer and an unchanged delivery system. Hydrogen could not be blended into the energy supply as simply, because the energy contained per unit of volume of hydrogen is only about one-third that of methane. Furthermore, the physical and chemical properties of hydrogen differ from methane, and this would make it awkward for the consumer because changes in his gas burners would be required [3].

The variation in energy content, or "heating value," of fuel gases has led to phrases to describe the gas such as high heating value, high Btu, low heating value, low Btu, and pipeline quality [1]. The various definitions are shown in Table 4.3. "Pipeline quality" indicates that all the properties of the gas approximate those of natural gas and that therefore the gas can be readily mixed with existing natural gas supplies. Low-heating-value gas often contains substantial quantities of carbon monoxide (CO), which has nearly the same volumetric heating value as hydrogen. Because this gas mixture also has a low energy content per unit weight, it is uneconomical to transport in pipelines; consequently, it is usually consumed close to the site of production [1].

The typical coal gasification process that yields a hydrogen product is depicted in Figure 4.1. This is one of the most likely processes to be used in the future (other process descriptions can be found in reference 1). To date, only pilot SNG plants are in opera-

---

*The abbreviation SNG originally meant "synthetic natural gas"; now, however, SNG is usually said to mean "substitute natural gas."

TABLE 4.3

Common Gas Nomenclature

| Common Name | Approximate Energy Content[a] (Btu/SCF[b]) |
|---|---|
| High heating value ⎫<br>High Btu ⎬<br>Pipeline quality ⎭ | 900–1,000 |
| Low heating value ⎱<br>Low Btu ⎰ | 100–500 |
| Methane ($CH_4$) | 1,000 |
| Hydrogen ($H_2$) | 325 |
| Carbon monoxide (CO) | 322 |

[a]Higher heating values.
[b]Standard cubic foot.

tion, although some plans are being made to construct commercial plants in the near future in the Four Corners area of New Mexico and in North Dakota [4]. Considerable water is consumed by these processes, both as a source of hydrogen and for cooling. Availability of water is an important constraint on the conversion of western coal resources.

## HYDROGEN FROM WATER BY ELECTROLYSIS

The electrolysis of water ($H_2O$) to produce both hydrogen and oxygen is demonstrated in every high-school chemistry course. This approach to hydrogen production has been well known for many years, and the hydrogen produced is quite pure [1, 3, 5-15]. However, until lately, little effort has been made to improve the energy efficiency or cost of the process. Relatively little hydrogen has been produced commercially by electrolytic processes,* because for decades when large quantities have been needed it has been more economical and more convenient to obtain hydrogen by reforming methane or oils [2].

---

*Total worldwide electrolytic production of hydrogen is only equivalent to about 3 percent of current U.S. total production of hydrogen.

FIGURE 4.1

Typical High-Temperature, Atmospheric-Pressure Coal Gasification Process

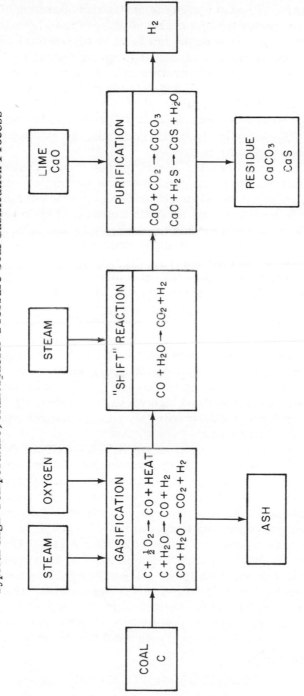

39

The reliability and simplicity of unattended operation of electrolysis equipment rather than its energy efficiency have been the prime design considerations, because the low level of commercial interest in this process did not provide economic incentive for the performance of innovative research and development. As a result, commercial electrolyzers have been small and relatively inefficient, and the technology has been rather stagnant. Commercially available electrolysis cells operate in the range of 60 to 70 percent efficiency [9]. The rest of the energy is dissipated in driving the electric current through the cell and appears as heat.

The need to provide electric power aboard manned and unmanned spacecraft has led to the development of fuel cells*—devices in which certain chemical reactions generate an electric current [16, 17]. The basic reaction is the catalytic oxidation of hydrogen rather than combustion; the end product is water, and some heat. The electrochemistry in electrolyzers and in fuel cells is essentially identical—indeed, the processes are basically the inverse of one another [3, 18]. As a result, in principle, it is possible to build a single device that can function either as an electrolyzer or a fuel cell, depending on whether it were operated in the "forward" or "backward" mode. Such a device is sometimes called a "reversible fuel cell." Although such a device might sound attractive for a hydrogen energy economy, a combined device is not very practical, because obtaining a very effective electrolyzer or a very effective fuel cell requires an engineering optimization of conflicting variables. Thus, a good fuel cell tends to be a poor electrolyzer and vice versa. Consequently, electrolyzers and fuel cells will almost certainly remain distinct entities in most practical applications.

In the last few years, increased research and development effort has been directed toward development of more efficient electrolyzers. It is generally agreed that improved electrolyzer efficiencies can be obtained by the use of higher temperatures and higher pressures of operation [3]. Separate experiments and demonstrations have reported operating temperatures of 2,000°F [3] and hydrogen and oxygen evolution pressures of 3,000 psi [13]. Unfortunately, however, as might be expected, operation at high pressures and temperatures has uncovered previously unexperienced materials problems, such as membrane and gasket degradation, which limit the useful lifetime of the cells. Electrochemists and metallurgists generally expect that the materials problems of electrolysis and fuel cells can be solved, not only on an experimental device scale but also on a commercial scale.

---

*Also silicon photovoltaic solar cells.

The efficiency of electrolysis is normally defined as the energy that ideally could be recovered by reoxidation of the hydrogen and oxygen coproducts to water, divided by the energy supplied to the electrolysis system in electrical form [3]; this definition is shown in Figure 4.2. Since electrolysis involves an endothermic chemical reaction (that is, one in which heat must be added), a perfect electrolyzer would consume both electricity and heat. Yet because this heat input is not counted in the usual definition of efficiency, the hydrogen product of a perfect electrolyzer would contain more energy than had been supplied in electrical form. Indeed, the maximum theoretical efficiency—using this definition—is 120 percent [3].* To date, in

FIGURE 4.2

Ideal Electrolysis Cell and Conventional Definition of
Electrolysis Efficiency

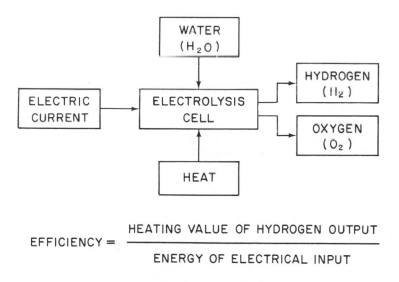

$$EFFICIENCY = \frac{HEATING\ VALUE\ OF\ HYDROGEN\ OUTPUT}{ENERGY\ OF\ ELECTRICAL\ INPUT}$$

Note: For present practical cells the direction of heat flow is reversed owing to electric current resistance losses in cell.

---

*Because electrolyzers and fuel cells share the same electrochemistry, the same considerations that allow a perfect electrolyzer to have an efficiency of 120 percent limit the efficiency of a perfect fuel cell to about 84 percent [3].

real-world electrolyzers, as opposed to ideal electrolyzers, resistance to the flow of electric current and other losses in the electrolysis cell still result in a net release, rather than an absorption, of heat. Such resistance would have to be nearly completely eliminated before an electrolysis cell could be used as a heat sink. However, should technology ever advance to that point, the heat input might be derived from the waste heat of the power plant that generated the electricity used in the electrolysis cell.

One of the most attractive attributes of electrolytic production of hydrogen is the small "natural" size of the practical cell. Because many cells would be cascaded and paralleled in a commercial installation, the plant output could be tailored to virtually any level of output without sacrificing important economies of scale. In Chapter 12 it will become apparent that this flexibility in size provides electrolysis with a versatility not readily obtained in other forms of hydrogen production.

Catalysts are such a necessary part of advanced, high-efficiency electrolytic cells [18] that the future availability of catalytic materials on a large scale is important. Some of the experimental cells use platinum or other rare noble metals for catalysts [18], although nickel is sometimes employed [9]. In commercial applications nickel compounds are the most likely future catalyst of choice. However, nickel is not a very abundant metal, and the United States currently imports about 70 percent of its needs—about 58 percent of it from Canada [19]. Because there are many other uses for nickel, such as in high-strength steels and other alloys, the availability of nickel has to be considered [19]. Depending on the level of electrolytic production of hydrogen, however, the availability of nickel for catalysts need not develop into a problem. For example, the Synthetic Fuels Panel report estimates that an annual production of about $15 \times 10^{12}$ SCF of hydrogen (the panel's estimate of the gap between domestic natural gas supply and demand in 1985)* would require only about 1,500 tons of nickel [9]— a small amount compared with the 1972 U.S. primary production of about 16,000 tons and world production of 700,000 tons [20]. However, production of hydrogen at a level to supplant natural gas would severely strain domestic nickel production.

## CLOSED-CYCLE THERMOCHEMICAL DECOMPOSITION OF WATER

If water is heated to a high temperature, such as 2,000°C (about 3,600°F), a small fraction (about 4 percent) of the molecules

---

*There is a numerical inconsistency in the panel's figures; they state $15 \times 10^{18}$ where they mean $15 \times 10^{12}$.

decompose into hydrogen and oxygen [21]. Such temperatures are far too high, and the quantities far too small, to lend encouragement to the concept of direct, single-step decomposition of water on a commercial scale [22, 23]. However, some ingenious multiple-step, closed-cycle chemical processes have been devised that may effectively use heat to decompose water on a commercial scale [3, 9, 21, 24–36].

The basic idea is that by several chemical reaction steps water can be broken into its hydrogen and oxygen constituents with all the other, intermediate reactants being continuously recycled. Figure 4.3 shows how the scheme can be viewed as splitting water by the application of heat and the equations of the Mark IX thermochemical water decomposition process [22, 29], developed at the Euratom facility at Ispra, Italy. Many other cycles are now being investigated.* Although the first cycles to be announced were devised by the combination of ingenuity and intuition, the search for usable cycles has been broadened and computerized at General Atomic [31]. The many possible combinations of chemical reactions are now being screened with respect to chemical and practical constraints.

Very high temperatures are usually required for one or more steps in the cycle—often in excess of 700°C (1,300°F). At present, only two (nonfossil-fueled) methods of sustaining such temperatures offer much promise—the high-temperature, gas-cooled nuclear reactor (HTGR) [35, 36] and highly focused solar radiation [37]. At the temperatures involved, much of the basic physical chemistry data necessary to evaluate the cycles fully have never been measured. Considerable laboratory research is needed to establish the thermodynamics and kinetics of the main reactions in even the most promising cycles, and approaches to the suppression of spurious parasitic chemical side reactions are needed as well. In addition, considerable further research is needed to understand and manage the materials problems that arise in attempting to contain the reactions without destructive corrosion of containment vessels and contamination of the chemical reactants. Bench tests have been performed on all steps for only a few cycles.

A potentially serious, but seldom mentioned, drawback to closed-cycle thermochemical production arises because the cycles are unlikely to be truly closed. Instead, there will be some loss of reactants, and, at the quantities of hydrogen production implied by

---

*By research groups at Euratom, General Atomic, General Electric, the Institute of Gas Technology, Lawrence Livermore Laboratory, Argonne National Laboratory, Oak Ridge National Laboratory, Los Alamos National Laboratory, and the University of Kentucky.

FIGURE 4.3

Schematic of Closed-Cycle Thermochemical Water Splitting
and Equations of the Euratom Mark IX Cycle

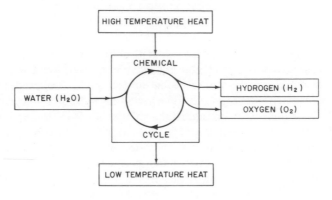

EURATOM MARK IX PROCESS

|  | TEMPERATURE |
|---|---|
| $6\ FeCl_2 + 8H_2O \longrightarrow 2\ Fe_3O_4 + 12\ HCl + 2H_2$ | (650°C) |
| $2\ Fe_3O_4 + 3\ Cl_2 + 12\ HCl \longrightarrow 6\ FeCl_3 + 6H_2O + O_2$ | (120°C) |
| $6\ FeCl_3 \longrightarrow 6\ FeCl_2 + 3\ Cl_2$ | (420°C) |

NET: $2H_2O \longrightarrow 2H_2 + O_2$

the hydrogen economy concept, even the loss of a fraction of a percent per cycle would release large amounts of reactants. Some of
the reactants proposed, such as mercury, cause undesirable environmental effects.

Why then, in view of the difficulties and uncertainties mentioned, is there attraction to the thermochemical decomposition of
water in preference to electrolysis of water? The answer lies in the
projected net energy efficiencies of the thermochemical processes.
Some hope to achieve a net efficiency of thermochemical water decomposition of about 70 percent. Researchers who have tried to estimate and deduct the energy expended in the chemical separations,
pumping, and so forth, have concluded that the upper limit of net energy efficiency will be about 55 percent [23]. Nevertheless, a net
efficiency of 55 percent is greater than the projected net efficiency of
36 percent for electrolysis (40 percent thermal efficiency for electricity generation from an HTGR cascaded with electrolysis at 90
percent efficiency).* Thus, achievement of a practical cyclic thermo-

*However, development of more advanced power generation cycles
may raise the efficiency of power generation, thereby making electrolysis a bit more competitive.

chemical decomposition process might make hydrogen production less energy-consuming, and possibly less expensive as well.

Because the financial rewards for achieving and patenting a process could be very substantial, * the interchange of thermochemical cycle information is considerably more guarded than in most other hydrogen economy subject areas. Perhaps more private-sector research and development money has now been devoted to this portion of the hydrogen economy than to any other.

## MIXED THERMAL/ELECTROLYTIC

Some hybrid cycles that combine thermochemical and electrolytic approaches are now being considered. Table 4.4 depicts a sulfur-based cycle being investigated at Westinghouse. Such approaches have received relatively little serious attention in the literature.

### TABLE 4.4

### Westinghouse Hybrid Cycle

| | |
|---|---|
| Electrolysis | $2\ H_2O + SO_2 \rightarrow H_2SO_4 + H_2$ |
| Thermal | $2\ H_2SO_4 \rightarrow 2\ H_2O + 2SO_2 + O_2$ |
| Net reaction | $2\ H_2O \rightarrow 2\ H_2 + O_2$ |

## HYDROGEN FROM THERMONUCLEAR FUSION

Poorly documented news reports have appeared in scientific magazines stating that KMS Fusion Corp. (in Ann Arbor, Michigan), the only private company to have made notable progress in the achievement of thermonuclear fusion with high-powered lasers, has a process by which hydrogen and/or methane could be made in a laser fusion reactor [38]. Details of the process have not been revealed because they are proprietary to KMS and its sponsor in the gas synthesis work, Texas Gas Transmission Co. [38]. It has been stated, however, that

---

*General Atomic in particular seems to have much to gain because development of a commercial process would almost certainly give a major sales boost to its newly commercialized HTGR—a reactor with no U.S. commercial competitors.

## TABLE 4.5

### Comparison of Hydrogen Production Alternatives

| Process | Advantage | Disadvantage |
| --- | --- | --- |
| Reformation of methane | Presently the cheapest method | Scant long-term potential as a source because of limitations on methane supply |
| Coal gasification | Cheapest and most secure near-term alternative to methane reformation; abundant coal reserves in United States | Ultimate limitation is exhaustion of the coal resource; requires large plant size |
| Electrolysis of water | Proven reliable technology; small unit plant size; well suited to all terrestrial solar energy collection approaches; oxygen coproduct, easily separated for possible use and economic credit improves the economics; improvements in efficiency quite likely; can produce hydrogen at high pressures, thereby eliminating the need for costly compression to pipeline pressures | High cost, lower net energy efficiency, possible resource limitation on catalysts |
| Thermo-chemical decomposition | Potentially most efficient nonfossil processes; not tied to fossil-fuel resources; possibly compatible with high-temperature, focused solar collectors | Not a proven technology; materials problems in containment; complex large unit plant size expected; expected release of potentially harmful chemicals |

the process uses the neutrons [39, 40] produced in the fusion reaction to produce hydrogen by "radiolytic disassociation" [40].

## COMPARISON OF HYDROGEN PRODUCTION PROCESSES

The advantages and disadvantages of the three major hydrogen production approaches are summarized in Table 4.5. A very important advantage of electrolysis, in spite of its expected higher cost and relative inefficiency, is its avoidance of dependence on any fossil energy resource (unlike coal gasification) and its ability to operate at temperatures much lower than thermochemical cycles. In addition, small, practical unit size and low voltage requirements of electrolysis suit it very well for integration into terrestrial solar energy collection systems, particularly photovoltaic, thermal-electric, ocean thermal gradient, and windpower.* The already very low efficiency (15 to 20 percent) of electric generation from geothermal sources [30] argues against subsequent further net energy loss through electrolytic production of hydrogen until very efficient electrolysis is developed. Because of the temperature of geothermal resources, there is little prospect of geothermal energy being used for thermochemical decomposition of water.

## REFERENCES

1. H. C. Hottel and J. B. Howard, New Energy Technology: Some Facts and Assessments (Cambridge, Mass.: Massachusetts Institute of Technology Press, 1971).
2. P. Meadows and J. A. DeCarlo, "Hydrogen," in Mineral Facts and Problems, Bureau of Mines Bulletin 650, 1970, pp. 97-110.
3. D. P. Gregory et al., "A Hydrogen-Energy System," American Gas Association, Alexandria, Va., August 1972.
4. H. R. Linden, "Is the Synthetic Fuels Option Credible?" paper presented at the Third National Energy Forum, Washington, D.C., May 15-16, 1975.
5. M. Steinberg, "A Review of Nuclear Sources of Non-Fossil Chemical Fuels," paper presented at the meeting of the American Chemical Society, Boston, Mass., April 9-14, 1972.

---

*In fact, some of these solar energy technologies may become viable only if wedded to hydrogen energy storage systems used to smooth out the variations in production.

6. R. L. Costa and P. G. Grimes, "Electrolysis as a Source of Hydrogen and Oxygen," Chemical Engineering Progress 63, no. 4 (April 1967): 56.

7. W. Juda and D. M. Moulton, "Cheap Hydrogen for Basic Chemicals," Chemical Engineering Progress 63, no. 4 (April 1967): 59-60.

8. N. C. Hallet, "Study, Cost, and System Analysis of Liquid Hydrogen Production," National Aeronautics and Space Administration Circular No. 73226, June 1968.

9. "Hydrogen and Other Synthetic Fuels," a summary of the work of the Synthetic Fuels Panel, prepared for the Federal Council on Science and Technology R&D Goals Study, September 1972.

10. W. A. Titterington and A. P. Fickett, "Electrolytic Hydrogen Fuel Production with Solid Polymer Electrolyte," Eighth Intersociety Energy Conversion Engineering Conference, 1973, pp. 574-79.

11. "Solid Electrolytes Offer Route to Hydrogen," Chemical and Engineering News, August 27, 1973, p. 15.

12. F. C. Jensen and F. H. Schubert, "Hydrogen Generation Through Static Feed Water Electrolysis," in Hydrogen Energy, ed. T. N. Veziroglu (New York: Plenum, 1975), pp. 425-40.

13. L. J. Nutall et al., "Hydrogen Generation by Solid Polymer Electrolyte Water Electrolysis," in Hydrogen Energy, ed. T. N. Viziroglu (New York: Plenum, 1975), pp. 441-56.

14. J. B. Laskin, "Electrolytic Hydrogen Generators," in Hydrogen Energy, ed. T. N. Veziroglu (New York: Plenum, 1975), pp. 405-16.

15. J. O'M. Bockris, "On Methods for the Large Scale Production of Hydrogen from Water," in Hydrogen Energy, ed. T. N. Veziroglu (New York: Plenum, 1975), pp. 371-404.

16. T. H. Maugh II, "Fuel Cells: Dispersed Generation of Electricity," Science 178 (December 22, 1972): 1273-74B.

17. A. J. Appleby, "Fuel Cells and Electrolyzers in the Hydrogen Economy," in Proceedings of the Cornell International Symposium and Workshop on the Hydrogen Economy, ed. S. Linke (Ithaca, N.Y.: Cornell University, April 1975), pp. 197-211.

18. E. E. Hughes et al., "Strategic Resources and National Security: An Initial Assessment," RADC-TR-75-54, Defense Advanced Projects Research Agency, Rome Air Development Center, Rome, N.Y., April 1975.

19. Statistical Abstract of the United States, 1975 (Washington, D.C.: Bureau of the Census, U.S. Department of Commerce, July 1974).

20. L. T. Blank et al., "A Hydrogen Energy Carrier," Systems Design Institute, National Aeronautics and Space Administration-American Society for Engineering Education, Johnson Space Center, Houston, Texas, 1973.

21. J. Funk and R. Reinstrom, "Energy Requirements in the Production of Hydrogen from Water," Industrial and Engineering Chemistry 5, no. 3 (July 1966): 336–42.

22. J. E. Funk and R. M. Reinstrom, "System Study of Hydrogen Generation by Thermal Energy," in Energy Depot Electrolysis Systems Study, Final Report TID 20441, Allison Division of General Motors Report EDR 3714 (Washington, D.C.: U.S. Atomic Energy Commission, June 1964), vol. 2, supp. A.

23. G. DeBeni, "Thermochemical Water Splitting with Nuclear Heat," in Proceedings of the Cornell International Symposium and Workshop on the Hydrogen Economy, ed. S. Linke (Ithaca, N.Y.: Cornell University, April 1975), pp. 113–21.

24. J. B. Pangborn, "Thermochemical Cracking of Water," in Proceedings of the Cornell International Symposium and Workshop on the Hydrogen Economy, ed. S. Linke (Ithaca, N.Y.: Cornell University, April 1975), pp. 128–30.

25. H. Barnert, "Thermochemical and Nuclear Technology for Nuclear Water Splitting," in Proceedings of the Cornell International Symposium and Workshop on the Hydrogen Economy, ed. S. Linke (Ithaca, N.Y.: Cornell University, April 1975), pp. 130–40.

26. B. M. Abraham and F. Schreiner, "A Low Temperature Thermal Process for the Decomposition of Water," Science 180 (June 1, 1973): 959–60, and "Low Temperature Thermal Decomposition of Water," Science 182 (December 28, 1973): 1372–73.

27. R. H. Wentorf, Jr., and R. E. Hanneman, "Thermochemical Hydrogen Generation," Science 185 (July 26, 1974): 311–19; disputations by R. Shinnar and M. A. Soliman, W. L. Conger, K. E. Cox, and R. H. Carty, "Thermochemical Hydrogen Generation Heat Requirements and Cost," Science 188 (June 6, 1975): 1036–37; reply by R. H. Wentorf and R. E. Hanneman, "Thermochemical Hydrogen Generation Heat Requirements and Cost," Science 188 (June 6, 1975): 1037–38.

28. J. E. Funk, "Thermodynamics of Thermochemical Hydrogen," in Proceedings of the Cornell International Symposium and Workshop on the Hydrogen Economy, ed. S. Linke (Ithaca, N.Y.: Cornell University, April 1975), pp. 122–27.

29. J. E. Funk et al., "Evaluation of Multi-Step Thermochemical Processes for the Production of Hydrogen from Water," in Hydrogen Energy, ed. T. N. Veziroglu (New York: Plenum, 1975), pp. 457–70.

30. G. DeBeni, "Considerations on Iron-Chlorine-Oxygen Reactions in Relation to Thermochemical Water-Splitting," in Hydrogen Energy, ed. T. N. Veziroglu (New York: Plenum, 1975), pp. 471–82.

31. J. B. Pangborn and J. C. Sharer, "Analysis of Thermochemical Water-Splitting Cycles," in Hydrogen Energy, ed. T. N. Veziroglu (New York: Plenum, 1975), pp. 499–516.

32.  J. L. Russell and J. T. Porter, "A Search for Thermochemical Water-Splitting Cycles," in Hydrogen Energy, ed. T. N. Veziroglu (New York: Plenum, 1975), pp. 517-32.

33.  R. G. Hickman et al., "Thermochemical Hydrogen Production Research at Lawrence Livermore Laboratory," in Hydrogen Energy, ed. T. N. Veziroglu (New York: Plenum, 1975), pp. 483-98.

34.  J. B. Pangborn and D. P. Gregory, "Nuclear Energy Requirements for Hydrogen Production from Water," Ninth Intersociety Energy Conversion Engineering Conference, 1974, pp. 400-05.

35.  H. Barnert and R. Schulten, "Nuclear Water-Splitting and High Temperature Reactors," in Hydrogen Energy, ed. T. N. Veziroglu (New York: Plenum, 1975), pp. 115-28.

36.  R. N. Quade and A. T. McMain, "Hydrogen Production with a High Temperature Gas-Cooled Reactor (HTGR)," in Hydrogen Energy, ed. T. N. Veziroglu (New York: Plenum, 1975), pp. 137-54.

37.  A. F. Hildebrandt and L. L. Vant-Hull, "A Tower-Top Point Focus Solar Energy Collector," in Hydrogen Energy, ed. T. N. Veziroglu (New York: Plenum, 1975), pp. 35-44.

38.  Russell D. O'Neal, Chairman of the Board and Chief Executive Officer, KMS Fusion, personal communication, 1975.

39.  "Laser Fusion: The Phoenix at KMS," Industrial Research, May 1975, pp. 30-32.

40.  U.S., House, Committee on Government Operations, Laser Fusion: A Solution to the Natural Gas Shortage?: Hearing, 94th Cong., 1st sess., June 3, 1975.

41.  E. E. Hughes et al., Control of Environmental Impacts from Advanced Energy Sources, Environmental Protection Agency Report EPA 600/2-74-002, March 1974.

# STORAGE OF HYDROGEN

## INTRODUCTION

Hydrogen must be stored to facilitate its distribution to end-use consumers and to provide an inventory to buffer variations in supply and demand. Gaseous storage in tanks appears likely to remain impractical except on a very small scale because high pressures are required and this necessitates the use of very heavy and robust tanks. However, there are other candidate storage mechanisms for gaseous hydrogen. Although hydrogen has the second lowest normal boiling point of any substance, 20.4°K (-253°C or -423°F),* it can also be stored as a cryogenic liquid. In addition, hydrogen can be stored in the form of metal hydrides, or by chemical combination to form other compounds. There is no single storage mechanism that would be best for all forms of hydrogen.

## STORAGE OF GASEOUS HYDROGEN

### Underground Storage

Gaseous hydrogen can be stored underground in three ways: in man-made or natural caverns, in depleted natural gas fields, or in suitable aquifers under an impervious rock cap [1]. Worldwide, all of these means to store gases containing hydrogen have been tried

---

*Absolute zero is 0°K (or equivalently -273°C, -460°F). Only helium has a lower normal boiling point, 4.2°K.

51

to some extent and, while many technical uncertainties remain, all will probably be used when the geologic and economic conditions are favorable.

A major difficulty in the use of underground caverns is the need to seal all openings so that the pressurized hydrogen gas cannot escape. In relatively nonporous geologic formations, most of the gas entered into storage can later be withdrawn and recovered [1]. This easy and nearly total recovery is not just an important economic advantage, it is also an advantage affecting the rate of injection and withdrawal [1].

Hydrogen gas can also be pumped underground into depleted natural gas fields to occupy the space formerly occupied by the natural gas [1]. Such a storage approach is limited by the relatively small void volume available per unit volume of rock (porosity) and the need for the gas to migrate through the small passages interconnecting the tiny voids (permeability). Thus, the resulting slow injection and withdrawal rates necessarily limit the application of this form of storage to situations that do not require rapid charge or withdrawal [1]. Although hydrogen is known for its ability to leak readily, the rate of loss of "town gas" (largely hydrogen) from a storage facility of this type in France has apparently been quite acceptable [1]. Helium, another very diffusive gas, was stored in depleted natural gas fields in the southwestern United States as part of a resource conservation effort for a number of years [2, 3] and thus provides related technological experience for this form of storage.

One drawback to this kind of storage is the requirement that a certain amount of hydrogen be irretrievably invested in initially charging the field, because by no means all of the hydrogen injected can later be withdrawn [1]. Another drawback affects quality control. Hydrogen withdrawal flushes out some of the residual natural gas in the field, so the hydrogen withdrawn is no longer pure; instead, the composition of the withdrawn gas is variable [1]. For some end-uses, however, such as combustion, this contamination presents few problems; but when hydrogen is used as a chemical, the contamination is at best a nuisance.

In some locations, porous rock aquifers are overlaid by nearly impervious cap rock, which is saturated by water, forming an effective gas seal. Hydrogen can be stored in the aquifer beneath this lid of rock and water [1]. Charging the aquifers with hydrogen displaces some of the water, but a layer of trapped water remains at the top of the aquifer. The seal that retains the hydrogen reportedly comes nearly entirely from the surface tension of the water in the cap rock [1]. For this reason, leakage is not a function of the molecular size of the stored gas, and this makes this form of hydrogen storage potentially practical.

Because of the large scale implied, underground gaseous hydrogen storage methods are probably more attractive to the utility industry than to other end-use sectors. In particular, electric utilities could use these approaches to the storage of hydrogen in load-leveling systems, and gas utilities could store hydrogen for seasonal load leveling.

### Pipeline Storage

Hydrogen gas can be stored in transmission pipelines whenever the maximum flow through the pipe is not being fully exploited [1]. In this "linepack" technique, the pipeline design pressure is maintained, but injections and withdrawals (and hence, "throughput") fall below the maximum possible. For very short-term storage, the pipeline may even be allowed to exceed normal operating pressure. The amount of storage that can be realized by the linepack technique depends heavily and in a complex way on the total system volume, pressure ratings, and other attributes of the actual system [1]. In early stages of hydrogen economy implementation, when demand would fall somewhat below that which could be supplied by the first pipelines, linepack might prove a useful interim storage method.

### STORAGE OF LIQUID HYDROGEN

Because liquid hydrogen is a very low-temperature liquid with properties rarely if ever encountered by the common man, it cannot be stored as casually as water or most familiar fuels [4–7]. At the normal boiling point of hydrogen, many materials in contact with it become brittle and contract from their dimensions at room temperature. Obviously liquid hydrogen containment vessels must be designed differently than the usual simple envelope that is sufficient to contain familiar fuels such as gasoline. The vessel must be very well insulated against the stray intrusion of heat—even that received in the form of radiation from surrounding objects at room temperature. Over the years, experimental laboratories, especially those concerned with cryogenic physical chemistry and solid-state physics, have accumulated considerable experience in the containment of liquid helium, the element with the lowest normal boiling point (4.2°K) and liquid nitrogen (77°K). Thus, because it is bracketed in temperature by liquid helium and liquid nitrogen, understanding of liquid hydrogen containment has benefited. Moreover, large research efforts to understand liquid hydrogen have been undertaken as part of the U.S. space program.

Small quantities of cryogenic liquids are normally stored in a double-walled vessel with vacuum between the two walls (dewar).* Dewars are identical in principle to the glass thermos bottle familiar to most households. Because the amount of spurious heat that can leak into a dewar is related to its surface area, the larger the ratio of volume to surface area, the slower the rate of boil-off loss. Some boil-off is inevitable, however, and, because the ratio of gas volume to liquid volume at 20°K is about 50 to 1, the storage vessel must be able to withstand a pressure buildup between withdrawals of liquid and also be equipped with a gas-venting device. Some vessels in use today can be safely sealed without venting for weeks at a time.

Because of the storage efficiency gained in large containers with a small surface-to-volume ratio (see Table 5.1), it has been possible to simplify greatly the construction principles of large tanks without encountering unacceptable boil-off losses. In the U.S. space program, which uses liquid hydrogen as a rocket propellant, tanks as large as 900,000 gallons have been constructed [4]. In these tanks the double walls are separated by "perlite" insulation about a foot thick and are evacuated [4, 6]. As Table 5.1 shows, boil-off rates for tanks this large are as low as 0.03 percent per day.

Storage tanks are not allowed to warm appreciably above 20°K because above that temperature there is a large amount of boil-off when liquid hydrogen is reintroduced into the warm tank, again chilling it down to liquid hydrogen temperature. Thus, unless a tank is in need of repair, some hydrogen is routinely left in the tank to maintain a low temperature. Flexible, vacuum-insulated transfer tubes are used during an addition or withdrawal transfer operation to a storage vessel. Tanks are maintained at a pressure slightly above ambient to prevent outside air from infiltrating the tank, because air immediately freezes into a solid composed of oxygen, nitrogen, and ice, and this clogs passages or forms an undesirable residue in the tank bottom. Moreover, because air taken from room temperature down to cryogenic temperatues contracts by a factor of about 700, a small inward leak can "cryopump" large quantities of air into the storage tank. A well-designed and correctly maintained transfer system surmounts this problem, but technical refinements necessary for storage and transfer pose barriers to the casual use of liquid hydrogen by untrained personnel.

At present, small liquid hydrogen storage vessels are expensive, and this poses an obstacle to the use of liquid hydrogen in modest quan-

---

*At 4.2°K and 20.4°K any residual trapped gas (such as air) freezes out, thereby improving the degree of vacuum initially established at room temperature.

TABLE 5.1

Storage Efficiency of Liquid Hydrogen Containers

| Capacity (gallons) | Class of Use | Boil-Off (percent per day) | Reference |
|---|---|---|---|
| 900,000 | Stationary | 0.03 | 4 |
| 500,000 | Stationary | 0.05 | 4 |
| 28,000 | Railcar delivery | 0.3 | 8 |
| 13,000 | Truck delivery | 0.5 | 8 |
| 260 | Mobile | 1.0 | 9 |
| 40 | Mobile | 2.0 | 9 |

tities—such as in automobiles. However, new design concepts, materials, and fabrication techniques are beginning to emerge that very likely will improve this situation immensely. Aluminum tanks enclosed in rigid, closed-cell plastic foam are an example of the promising new directions.

The volumetric energy density of hydrogen is only about one-third that of conventional hydrocarbon fuels such as gasoline and jet fuel. Consequently, to deliver the same energy, storage tanks must contain about three times as much liquid and, hence, are considerably larger than those now used for fuel storage.

## LIQUEFACTION FOR STORAGE

Liquefaction of hydrogen is a complicated exercise in cryogenic engineering [10, 11]. Today, worldwide, there are very few commercial-scale liquid hydrogen plants, and the bulk of the output still goes to the U.S. space program. The Linde Division of Union Carbide Corporation has been the most active producer of cryogenic liquids to contribute to the hydrogen energy economy concept. Linde plants obtain their hydrogen from the reformation of methane (natural gas) [12].

To liquefy hydrogen, energy must be withdrawn from the gas to condense it [10]. It is not a simple matter to cool the hydrogen, and the complicated process depends heavily on the proper operation of efficient heat transfer devices at cryogenic temperatures [12]. Before liquefaction begins, and during the process, the hydrogen must be kept very pure because every contaminant except helium (an unlikely contaminant) will condense and freeze in the piping, thereby clogging the piping and also rendering the heat transfer surfaces less efficient [12]. Linde reportedly achieves less than 1 ppm total impurity before liquefaction begins [12].* Scant improvement is expected in the energy efficiency of liquefaction because it is now being achieved at about 40 percent of the theoretical (but never attainable) "ideal" efficiency [11]. For cryogenic liquids, this is good performance. The efficiency of liquefaction (using current technology) is about 77 percent if efficiency is defined as the energy contained in the heat of combustion of the liquid hydrogen, divided by the sum of the heat of combustion of the gaseous starting hydrogen and the electrical energy needed to achieve liquefaction [11]. This efficiency figure includes the effort required in a physical phase transition unique to hydrogen—the "ortho-to-para" conversion [13].† Efficient liquefaction is almost certainly limited to large-scale plants, although less efficient liquefaction might be tolerated for small-scale uses. Small (for example, 50 liters per day) liquefaction units that are not very energy efficient are commercially available today.

Liquid hydrogen is shipped from commercial liquefaction plants in large dewars mounted on semitrailer trucks (capacity about 13,000 gallons) or on railroad cars (capacity about 28,000 gallons). The trucks are used for deliveries within about a 1,000-mile range, and the railcars are sometimes used to deliver hydrogen across the continent [12]. The boil-off rates of these mobile storage containers is sufficiently low (see Table 5.1) that hydrogen gas is not vented en route but rather is allowed to build up pressure in the tank [12]. Such pressurization is useful because it provides a driving force for

---

*As a result of this purity, industries that need very pure gaseous hydrogen for chemical uses (for example, the semiconductor industry) purchase liquid hydrogen and then gasify it to obtain their hydrogen chemical reagent. This use of hydrogen, while important for what it achieves, currently consumes only a small portion of total liquid hydrogen production.

†The hydrogen molecule can exist in either of two molecular "spin" states called ortho and para. Both states are present when hydrogen is liquefied (75 percent ortho and 25 percent para). In the liquid state the para spin state is stable but the ortho is not. Left alone, the hydrogen would gradually convert almost entirely to the para state. This transition, however, releases enough energy to cause substantial boil-off of the hydrogen. Thus, unless this physical transition is driven to completion artificially, with the heat released removed by refrigeration, the hydrogen would boil off during storage as the natural ortho-to-para conversion proceeded. A catalyst is used in this conversion.

hydrogen withdrawal, although in practice the vaporized hydrogen is subsequently lost.

## STORAGE IN THE FORM OF METAL HYDRIDES

Surprisingly, as shown in Table 5.2, more hydrogen atoms can be packed into some metal hydrides than into the same volume of liquid hydrogen [14]. Not all metal hydrides pack hydrogen so effectively, but many do (for example, magnesium hydride and iron-titanium hydride).

Storage in a metal hydride is a physical-chemical process involving the diffusion of hydrogen atoms through the crystal structure of a solid metal or alloy where they react to form chemical compounds [15]. As shown schematically in Figure 5.1, the hydrogen atoms take up residence between the atoms composing the crystal of the host material. Specific metal hydrides absorb only well-defined maximum amounts of hydrogen; accordingly, the metal hydride formula (for example, $MgH_2$, $LaNi_5H_6$, $FeTiH_2$) is written in a manner that represents the maximum hydrogen concentrations attainable.

Because the rate of formation and decomposition of a metal hydride depends on the rate that hydrogen can diffuse from the outer surface of the metal inward to a vacant site in the crystal structure, pieces of metal with a surface-to-volume ratio as large as possible are used in applications of hydrides. Normally small metal granules or a metal powder are used as the host.

The formation of most metal hydrides is exothermic [14, 16]; that is, heat is released as the conversion from metal to metal hydride proceeds. Consequently, during the formation process, heat must be removed from the "bed" of particles by heat transfer devices

TABLE 5.2

Relative Hydrogen Density of Various Substances

|  | Density (grams/cm$^3$) | Weight Percentage Hydrogen | Number of Hydrogen Atoms/cm$^3$ |
|---|---|---|---|
| $LH_2$ (liquid hydrogen) | 0.07 | 100.0 | $4.2 \times 10^{22}$ |
| $NH_3$ (ammonia) | 0.6 | 17.7 | $6.5 \times 10^{22}$ |
| LiH | 0.8 | 12.7 | $5.3 \times 10^{22}$ |
| $MgH_2$ | 1.4 | 7.6 | $6.7 \times 10^{22}$ |
| $TiH_2$ | 3.8 | 4.0 | $9.1 \times 10^{22}$ |

FIGURE 5.1

Interstitial Sites for Hydrogen in a Crystal Lattice

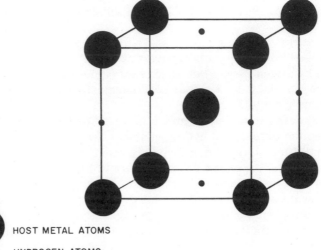

HOST METAL ATOMS

●     HYDROGEN ATOMS

embedded in it [14]. To release gaseous hydrogen it is necessary
to reverse the process of hydride formation, which means that the
same quantity of heat withdrawn during formation of the hydride must
be added to it. The rates at which hydrides form or decompose are
unique to each hydride substance. Curves showing the amount of hydrogen absorbed at different temperatures and pressures in $Mg_2NiH_2$
and FeTiH, two of the hydrides thought to be most useful in practical
devices, are shown in Figures 5.2 and 5.3.

Since metal hydrides are exothermic on formation and endothermic on decomposition, it has been suggested that if a metal hydride were used as the fuel tank of an automobile, the waste heat of
the engine rejected in its exhaust could be used to extract the hydrogen [17-19]. This would utilize energy normally wasted into the atmosphere. Conversely, on hydride formation, the same amount of
energy is released, and it might be possible to collect it and put it
to a beneficial use [19]. Thus, in effect, some of the waste energy
from hydride-equipped vehicles might be conserved and put to use.
This last concept of energy collection is probably not really viable,
however, if the depleted fuel tank is recharged at hundreds of thou-
sands of individual service stations.

Metal hydrides offer several potential advantages over the storage of hydrogen in liquid form. First, because the hydrogen is handled in gaseous form, the energy expenditure associated with liquefaction is saved. Second, while a leak in a liquid hydrogen tank re-

FIGURE 5.2

Pressure-Composition Isotherms for the $Mg_2Ni$-H Metal Hydride System

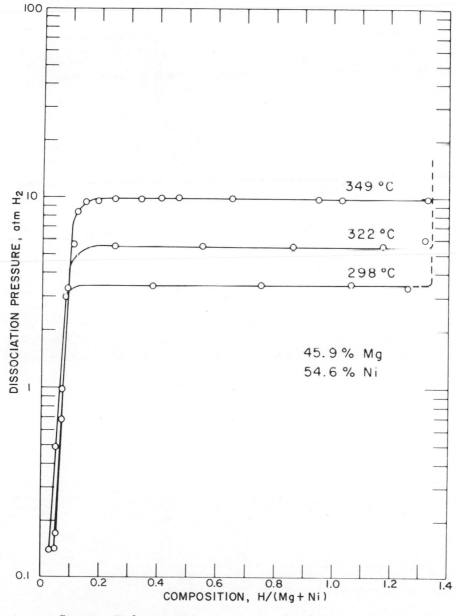

Source: Reference 14.

FIGURE 5.3

Pressure-Composition Isotherms for the FeTi-H Metal Hydride System

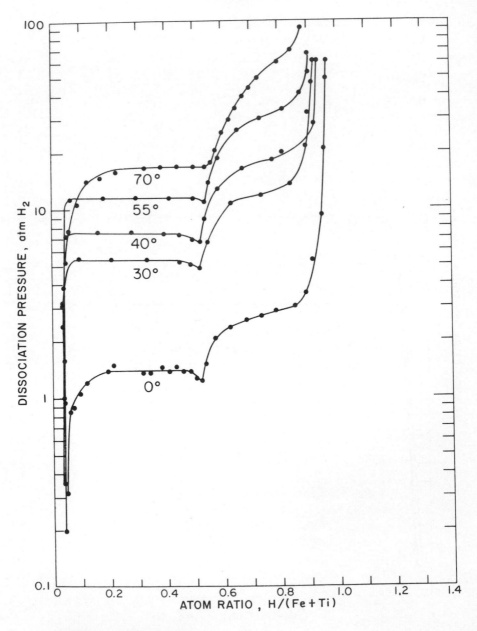

Source: Reference 18.

sults in a spill and immediate evaporation and dissipation of the hydrogen, perhaps with danger to people (see Chapter 8), a rupture in the vessel containing the metal hydride particles would not result in the release of hydrogen unless heat were applied; this offers safety against unwanted fuel combustion. Third, as already noted, the packing density of the hydrogen atoms can be higher than in the liquid.

There are also some drawbacks to the use of metal hydrides. First, impurities in the hydrogen gas, especially sulfur and oxygen, "poison" the hydride bed, thereby interfering with the hydride formation. This phenomenon is not now understood [15]. Second, metal hydrides are brittle materials, and repeated charge-discharge storage cycles tend to fracture the metal particles. As they become ever smaller, the particles tend to settle, pack, and cake. This slows the passage of hydrogen gas among the particles and thus reduces the reaction rates (in spite of the increased net surface to volume ratio). Third, metal hydrides are far heavier than liquid hydrogen (even one of the lightest, $MgH_2$ is about 13 times heavier for the same quantity of hydrogen stored). This makes the use of metal hydrides more unwieldy for mobile storage than for stationary storage.

Considerable research is now under way, especially at Brookhaven National Laboratory, to characterize further the properties of candidate metal hydride compounds and to demonstrate storage systems on a practical scale.

## STORAGE IN CHEMICAL COMPOUNDS

There are many other chemical compounds in which hydrogen can be stored and transported. These compounds can then be consumed directly or first decomposed to yield the hydrogen. Candidate compounds include ammonia, $NH_3$, hydrazine, $N_2H_4$, boranes, silanes, and synthetic carbons containing compounds such as alcohols. Of these, ammonia is possibly the most interesting because the only other element required, nitrogen, is the major constituent (80 percent) of air.

Some people would therefore include these compounds as part of the hydrogen energy economy concept. Indeed, at hydrogen energy symposia, papers on methanol (methyl alcohol) and ammonia are sometimes presented. However, to limit the scope of this study we have exluded these chemical compounds from consideration, because, in our opinion, the main thrust of the hydrogen energy concept begins to fade substantially in these cases.*

*We recognize, however, that the inclusion of metal hydrides and the exclusion of ammonia—which can be viewed as a nitrogen hydride—is somewhat arbitrary.

## REFERENCES

1.  D. P. Gregory et al., "A Hydrogen-Energy System," American Gas Association, Alexandria, Va., August 1972.

2.  W. D. Metz, "Helium Conservation Program: Casting It to the Winds," Science 183, no. 4120 (January 11, 1974): 59–63.

3.  C. A. Price, "The Helium Conservation Program of the Department of the Interior," in Patient Earth, ed. J. Harte and R. H. Socolow (New York: Holt, Rinehart and Winston, 1971), pp. 70–83.

4.  J. R. Bartlit, F. J. Edeskuty, and K. D. Williamson, Jr., "Experience in Handling, Transport, and Storage of Liquid Hydrogen —the Recyclable Fuel," Seventh Intersociety Energy Conversion Engineering Conference, 1972, pp. 1312–15.

5.  F. A. Martin, "The Safe Distribution and Handling of Hydrogen for Commercial Application," Seventh Intersociety Energy Conversion Engineering Conference, 1972, pp. 1335–41.

6.  F. J. Edeskuty and K. D. Williamson, Jr., "Storage and Handling of Cryogens," in Advances in Cryogenic Engineering, vol. 17, ed. K. D. Timmerhaus (New York: Plenum, 1972), pp. 1312–49.

7.  J. R. Bartlit, "Liquid Hydrogen Handling, Transport and Storage," in Proceedings of the Cornell International Symposium and Workshop on the Hydrogen Economy, ed. S. Linke (Ithaca, N.Y.: Cornell University, April 1975), pp. 95–101.

8.  John E. Johnson, Linde Division, Union Carbide Corporation, New York, personal communication, 1975.

9.  L. W. Jones, "Liquid Hydrogen as a Fuel for Motor Vehicles: A Comparison with Other Systems," Seventh Intersociety Energy Conversion Engineering Conference, 1972, pp. 1364–65.

10.  R. F. Barron, "Liquefaction Cycles for Cryogens," in Advances in Cryogenic Engineering, vol. 17, ed. K. D. Timmerhaus (New York: Plenum, 1972), pp. 20–36.

11.  W. R. Parrish and R. O. Voth, "Cost and Availability of Hydrogen," in Selected Topics on Hydrogen Fuel, ed. J. Hord, Report NBS IR 75–803 (Boulder, Colo.: Cryogenics Division, National Bureau of Standards, January 1975), pp. 1–25.

12.  T. J. Lewanski, Linde Division, Union Carbide Corporation, Fontana, Calif., personal communication, 1974.

13.  L. T. Blank et al., "A Hydrogen Energy Carrier," Systems Design Institute, National Aeronautics and Space Administration-American Society for Engineering Education, Johnson Space Center, Houston, Texas, 1973.

14.  R. D. Wiswall, "Hydrogen Storage Via Hydrides," in Proceedings of the Cornell International Symposium and Workshop on the Hydrogen Economy, ed. S. Linke (Ithaca, N.Y.: Cornell University, April 1975), pp. 102–09.

15.  Richard Wiswall, Brookhaven National Laboratory, Upton, N.Y., personal communication, 1975.

16.  R. H. Wiswall and J. J. Reilly, "Metal Hydrides for Energy Storage," Seventh Intersociety Energy Conversion Engineering Conference, 1972, pp. 1342-48.

17.  R. E. Billings, "Hydrogen Storage in Automobiles Using Cryogenics and Metal Hydrides," in Hydrogen Energy, ed. T. N. Veziroglu (New York: Plenum, 1975), pp. 791-801.

18.  C. H. Waide, J. J. Reilly, and R. H. Wiswall, "The Application of Metal Hydrides to Ground Transportation," in Hydrogen Energy, ed. T. N. Veziroglu (New York: Plenum, 1975), pp. 779-90.

19.  A. L. Austin, "A Survey of Hydrogen's Potential as a Vehicular Fuel," University of California Radiation Laboratory—51228, Livermore, Calif., June 1972.

# 6

## DISTRIBUTION OF HYDROGEN

### DISTRIBUTION IN GASEOUS FORMS

Gaseous hydrogen has long been distributed in small quantities at high pressure (2,000 psi) in robust steel cylinders. However, because the most commonly sized cylinders are very heavy (about 160 lbs) and contain very little hydrogen (only about 240 SCF or 1.3 lbs), this form of distribution is impractical for major use in a hydrogen energy economy.

By far the most practical form of large-volume hydrogen distribution is through pipelines in a manner akin to the present distribution of natural gas [1, 2].* Natural gas and hydrogen, however, are different in physical properties and affect materials differently [2-8]. Consequently, considerable attention must be given to the details of pipeline transmission. The natural gas industry, anticipating a worsening shortage of methane, has shown considerable interest in the distribution of gaseous hydrogen because it is the logical alternative to methane for use in homes, industry, and commerce. † The volumetric heating value of hydrogen is only about one-third that of methane; thus, for the same volumetric flow, a pipeline could deliver only about one-third as much energy in the form of hydrogen as it

---

*Although it is probably more romantic than practical, large-volume distribution in balloons and blimps might be useful to deliver relatively small amounts to remote locations distant from highways or railways where demand could not justify a pipeline.

†Accordingly, the American Gas Association has funded considerable study of hydrogen for utilities—most of it performed at the Institute of Gas Technology (IGT).

could in the form of methane. However, this disadvantage of hydrogen is offset by its lower density and viscosity (that is, it flows through the pipe with less frictional drag than does methane). On this premise, it was thought that nearly the same amount of energy might be transmitted feasibly either as methane or hydrogen [1, 2, 9-12]. Until 1974, therefore, when the first hydrogen pipeline optimizations were published [13-15], it was commonly believed that existing natural gas transmission pipelines could be converted to hydrogen use merely by the installation of more closely spaced and higher-capacity compression stations.

However, recent optimization studies have shown that a pipeline optimized to transmit hydrogen would deliver only about 0.6 to 0.7 times as much energy as the same pipeline operating at the same pressure but optimized for natural gas [14, 15], and hence larger diameter pipelines are implied. Cost studies have indicated that hydrogen pipelines should be operated at pressures conventionally used in natural gas pipelines (750 psi) or higher. If the hydrogen gas is produced at low pressure and must therefore be greatly compressed for use in the pipeline, large energy and cost penalties must be paid [16].* Although this finding somewhat diminishes the attractiveness of a conversion from methane to hydrogen, there are other more serious drawbacks.

The problem of hydrogen embrittlement of metals ranks first among the other drawbacks. There are two important classes of hydrogen embrittlement of metal [7, 8]: internal hydrogen reactions (for example, hydrogen combining chemically with the carbon in steel to form microscopic bubbles of methane, $CH_4$); and "environment" embrittlement. However, in transmission pipelines, only hydrogen environment embrittlement causes great concern [5, 7, 8]—primarily for reasons of system integrity and safety. The hydrogen environment embrittlement phenomenon is most pronounced at high pressures, high-purity hydrogen, and near room temperatures [7, 8]. The effect is manifested as the formation and propagation of cracks through a stressed metal or alloy until fracture occurs [3-8]. In effect, a normally ductile metal becomes brittle. This kind of embrittlement initiates essentially instantaneously (in less than a second) when the hydrogen and metal are exposed to each other; the effect ceases just as rapidly when the metal is removed from the hydrogen environment [7]. Although the onset of the effect is instantaneous, failure is not. Removal of the metal from the hydrogen environment does not remove the cracks; it merely removes the susceptibility. Although the actual physical mechanisms involved are not well understood, the occurrence of the phenomenon is well documented [3-8].

―――――――――――――――

*The same holds true for SNG.

Small amounts of some impurities in the hydrogen, notably oxygen, suppress embrittlement [7]. The intensity of the effect varies greatly from metal to metal, but high-strength steel alloys are highly susceptible, as are ordinary carbon steels [5]. There is hope that hydrogen environment embrittlement can be controlled by the use of protective plating with a less susceptible metal or through the introduction of controlled impurities [7]. Because hydrogen diffuses readily through plastics, insertion of a plastic liner or application of a plastic coating would not prevent embrittlement.

Metallurgists and engineers expect that a better understanding of hydrogen environment embrittlement can facilitate workable design and materials selection guidelines and fabrication codes that would guarantee sufficient compensatory strength. However, to the extent that more sophisticated configurations or materials were required, such design would probably consume more material and labor and thus result in increased fabrication costs. It is possible that the hydrogen normally transmitted would not be pure enough to cause embrittlement; thus, in practical application hydrogen environment embrittlement may be much less a problem than laboratory experiments suggest.

A pipeline about 130 miles long has been distributing hydrogen in Germany for nearly 30 years with no apparent problems of hydrogen environment embrittlement [2, 17-19]. This example is frequently cited in literature on the hydrogen economy as a counter to suggestions that embrittlement is a serious problem. Unfortunately, the example is not fully apt. In the first place, the purity of the hydrogen always remains unspecified, and the pressures involved are only about 150 psi [2] compared with the 750 to 1,000 psi contemplated for hydrogen trunk pipelines [15]. Thus, while this experience serves as a valuable lesson, it is only a starting point for the analysis that must precede use of a pipeline to transmit hydrogen.

It is frequently asserted that the conversion of methane transmission pipelines to hydrogen would save this large capital investment from obsolescence. However, a number of considerations cast doubt on this argument. Figure 6.1 shows that the trunk natural gas pipeline network fans out from the major natural gas-producing areas largely concentrated in the Gulf Coast region. These either terminate or taper in capacity with increasing distance from the gas fields as consumption lowers the quantity left to transmit. There is no natural reason to concentrate major hydrogen-producing facilities in the Gulf Coast region—least of all hydrogen from coal gasification, a likely first source of hydrogen. Thus, because the existing trunk pipelines generally do not link the correct places and taper in the inappropriate direction, the argument that conversion to hydrogen would protect those investments seems overstated. At best, it appears that only some trunk pipelines would remain useful.

FIGURE 5.1

Natural Gas Transmission Pipelines in the United States

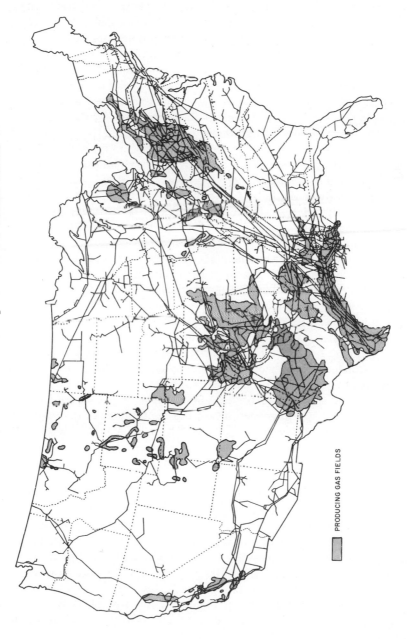

PRODUCING GAS FIELDS

Note that most pipelines originate in the Gulf Coast and the Southwest.
Source: Reference 1.

The possible leakage of hydrogen from pipelines is important
to consider. The volumetric loss rate of hydrogen through factures,
leaky seals, or small corrosion holes is about three times that of
methane, but the leakage of energy is nearly equal for both gases because of their relative volumetric energy densities [2]. In a confined
space, concentrations would build toward the flammable limit four
times as fast for hydrogen as for methane [2]. The typical leakage
rate of methane from existing gas system pipelines is between 100,000
and 400,000 SCF per mile per year. Old and leaky methane distribution pipelines have sometimes been rejuvenated by the insertion of a
plastic inner sleeve. Tests have shown that plastic pipelines leak far
less, about 80 SCF per mile per year, than the typical pipelines cited
above. Although hydrogen leaks through various plastics from 5 to 90
times as fast as methane [2], the total loss of the hydrogen from
plastic pipelines would be far less than presently accepted for methane.
Consequently, at least from the standpoint of leakage, most of the
existing natural gas pipelines at the local level should be suitable for
hydrogen. However, from the standpoint of capacity, the existing local pipelines may not be adequate because they would need to convey
about three times the volume of hydrogen as they now convey for
methane. If the pipeline pressure were raised to increase the gas
flow, then the question of hydrogen leakage would have to be reeval-
uated. The triple increase in volume flow would affect the operation
of gas meters. Most of the present gas-metering systems, especially
at the level of the dwelling unit, could not cope with this added flow
[2]. Consequently, most meters would have to be altered or replaced
—at considerable trouble and expense.

Thus, although at first sight it might be presumed that a switch
to hydrogen could preserve intact the investment in the present natural gas distribution system, in reality matters are not quite that
simple.


### DISTRIBUTION IN LIQUID FORM

The physical properties and handling characteristics of liquid
hydrogen are discussed in Chapters 4, 5, and 8. Today, most liquid
hydrogen is consumed by the U.S. space exploration program. Liquid
hydrogen is routinely transported by special railroad tank cars and
highway trucks to vehicle launch sites, as well as to sites of space
research and development, production, and testing [20, 21]. The
volumes distributed by railcar and highway truck* are limited by legal

---

*A representative of the Linde Division of Union Carbide Corpor-
ation reports that truck drivers who load and convey the liquid hydro-

size restrictions, but the weight limits have not been reached because of hydrogen's extremely low density [21].

It is technically possible to distribute liquid hydrogen through an extremely well-insulated cryogenic pipeline. However, the costs of such a pipeline are very high [23]; * such pipelines are economically practical for only very short distances and very high hydrogen flow rates. Consequently, in the early implementation of any aspect of a hydrogen energy economy requiring liquid hydrogen, delivery in batches by truck or train is most likely. Indeed, even for hydrogen shipments to Cape Canaveral for space launches, batch delivery is employed [24]. In the future, it generally would prove most economical to transport gaseous hydrogen by pipeline from its point of production to its point of liquefaction and then to distribute the liquid by truck or train. The sole exception may be in the commercial aviation sector. Near an airport it might prove advantageous (and even necessary for public safety reasons) to move liquid hydrogen underground a mile or so from the liquefaction plant or storage areas to the aircraft fueling stations. The high cost of such pipelines would be a strong incentive for adoption of a single central refueling terminal (as at Dulles Airport serving Washington, D.C.) instead of refueling at each departure gate as is now done for kerosene-type jet fuels.

## DISTRIBUTION IN METAL HYDRIDE FORM

Although it would be possible to distribute "charged" metal hydride beds and return "empty" beds to and from a central recharging point, the metal bed is heavy compared with the weight (and hence energy) of the hydrogen contained. Thus, in general, for reasons of weight, metal hydride beds cannot be expected to be used as a form of hydrogen distribution. Instead, when hydrides are used for hydrogen storage the hydrogen will be conveyed to the bed by pipeline.

However, in the automotive sector, metal hydrides might possibly be moved short distances to the user (although this would be limited by their heavy weight). For example, a depleted hydride bed from a car might be physically exchanged for a charged bed at the filling station. At the filling station, purified hydrogen could be used to recharge the bed, but it might turn out that economies of scale in hydrogen purification and hydride recharge would require a large re-

---

gen receive about three days' special instruction but are not otherwise unusual in education or background [22].

*About $100 per linear foot (or more than $500,000 per mile) for a pipe 5 inches in diameter [23].

gional hydride recharge center. In that event, hydride beds might be transported small distances between the center and filling stations, but this would probably prove too expensive to be viable.

Another approach to the use of metal hydrides in automobiles is to have a bed permanently located in the car that is charged by gas delivered by pipeline either to the home or to a "filling station."*

## SUMMARY

The most economical and convenient future way to distribute vast quantities of hydrogen is likely to be by gas pipeline. When needed, liquefaction could then be accomplished near the end-use, and most metal hydrides would be charged in place rather than transported. Even distribution by pipeline has a major economic hurdle to overcome: unless the hydrogen is produced at high pressure directly, the cost of pumping the hydrogen up to pipeline pressure will be high [16]. This will pose a large barrier to the implementation of nearly all facets of the hydrogen energy economy concept.

## REFERENCES

1. D. P. Gregory, "The Hydrogen Economy," Scientific American, January 1973, pp. 13-21.

2. D. P. Gregory et al., "A Hydrogen-Energy System," American Gas Association, Alexandria, Va., August 1972.

3. R. P. Jewett et al., Hydrogen Environment Embrittlement of Metals, National Aeronautics and Space Administration Report CR-2163, March 1973.

4. H. H. Johnson and A. J. Kumnick, "Hydrogen and the Integrity of Structural Alloys," in Hydrogen Energy, ed. T. N. Veziroglu (New York: Plenum, 1975), pp. 1043-55.

5. A. W. Thompson, "Structural Materials Use in a Hydrogen Economy," paper 37c, presented at the Annual Meeting, American Institute of Chemical Engineers, Washington, D.C., December 3, 1974.

6. H. H. Johnson, "Hydrogen Embrittlement," Science 179 (January 19, 1973): 228.

---

*High-purity hydrogen is important to the proper functioning and longevity of the hydride bed. Because the hydrogen obtained in the home from a common pipeline is unlikely to be reliably pure, we question the realism of this approach.

7. W. T. Chandler and R. J. Walter, "Hydrogen Environment Embrittlement of Metals and Its Control," in Hydrogen Energy, ed. T. N. Veziroglu (New York: Plenum, 1975), pp. 1057-78.

8. A. J. Kumnick, "Hydrogen Embrittlement," in Proceedings of the Cornell International Symposium and Workshop on the Hydrogen Economy, ed. S. Linke (Ithaca, N.Y.: Cornell University, April 1975), pp. 422-23.

9. D. P. Gregory, "A New Concept in Energy Transmission," Public Utilities Fortnightly, February 3, 1972, pp. 3-11.

10. D. P. Gregory and J. Wurm, "Production and Distribution of Hydrogen as a Universal Fuel," Seventh Intersociety Energy Conversion Engineering Conference, 1972, pp. 1329-34.

11. "Hydrogen and Other Synthetic Fuels," a summary of the work of the Synthetic Fuels Panel, prepared for the Federal Council on Science and Technology R&D Goals Study, September 1972.

12. L. T. Blank et al., "A Hydrogen Energy Carrier," Systems Design Institute, National Aeronautics and Space Administration—American Society for Engineering Education, Johnson Space Center, Houston, Texas, 1973.

13. G. Beghi et al., "Economics of Pipeline Transport for Hydrogen and Oxygen," in Hydrogen Energy, ed. T. N. Veziroglu (New York: Plenum, 1975), pp. 545-60.

14. R. A. Reynolds and W. L. Slager, "Pipeline Transmission of Hydrogen," in Hydrogen Energy, ed. T. N. Veziroglu (New York: Plenum, 1975), pp. 533-43.

15. A. Konopka and J. Wurm, "Transmission of Gaseous Hydrogen," Ninth Intersociety Energy Conversion Engineering Conference, 1974, pp. 405-12.

16. D. P. Gregory, "The Hydrogen Economy in Perspective," in Proceedings of the Cornell International Symposium and Workshop on the Hydrogen Economy, ed. S. Linke (Ithaca, N.Y.: Cornell University, April 1975), pp. 17-26.

17. C. Isting, "Experience with a Hydrogen Pipeline Network," Chemische Werke Büls AG, Marl, Federal Republic of Germany (undated).

18. C. Marchetti, "Hydrogen, Master Key to the Energy Market," Euro-Spectra 10, no. 4 (December 1971): 117-29.

19. C. Isting, "Pipelines Now Play Important Role in Petrochemical Transport," World Petroleum 41 (April 1970): 38-44.

20. J. R. Bartlit, F. J. Edeskuty, and K. D. Williamson, Jr., "Experience in Handling, Transport, and Storage of Liquid Hydrogen —the Recyclable Fuel," Seventh Intersociety Energy Conversion Engineering Conference, 1972, pp. 1312-15.

21. F. A. Martin, "The Safe Distribution and Handling of Hydrogen for Commercial Application," Seventh Intersociety Energy Conversion Engineering Conference, 1972, pp. 1335-41.

22.  John E. Johnson, Linde Division, Union Carbide Corporation, New York, personal communication, 1975.

23.  J. Hord, "Cryogenic $H_2$ and National Energy Needs," paper N-1 presented at the Cryogenic Engineering Conference, Atlanta, Ga., August 8-10, 1973.

24.  C. R. Dyer, M. Z. Sincoff, and P. D. Cribbins, eds., "The Energy Dilemma and Its Impact on Air Transportation," National Aeronautics and Space Administration-American Society for Engineering Education, Langley Research Center, Hampton, Va., 1973.

## VERSATILITY OF APPLICATION

Hydrogen could be used in nearly every energy application that relies on fuel combustion (oxidation). In addition, hydrogen offers opportunities for unconventional energy and chemical technologies— such as fuel cells and flameless catalytic burners. The technological options offered by hydrogen are the subject of this chapter.

## ENERGY UTILITIES

### Gas Utilities

Methane, the major component of natural gas, is easily distrib-uted in pipelines and stored in simple tanks and underground in geo-logical strata. Moreover, since it is relatively clean-burning and easy to control, it is a most favored fuel with rapidly increasing de-mand, not only in residences but in commercial establishments and in industry. Unfortunately, as shown in Figures 7.1 and 7.2, reserves of methane have been falling steadily as the soaring demand depletes producing gas fields [1]. Meanwhile, price regulation by the Federal Power Commission of natural gas shipped interstate has severely re-duced the incentive of the gas producer to discover and develop more reserves. Already some demand for natural gas is not being met be-cause domestic production [2] is inadequate. Some methane, includ-ing its liquefied form, is imported from other nations. The natural gas industry, recognizing that its survival is at stake, is turning to coal gasification to produce a synthetic methane.

FIGURE 7.1

U.S. Natural Gas Consumption

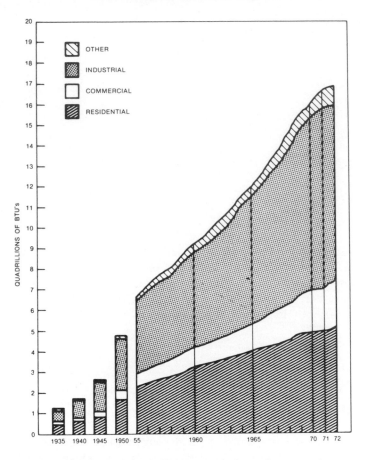

Source: Project independence report, reference 1.


For residences and commercial establishments, hydrogen is a potential substitute for natural gas. It could be used in a manner analogous to the use of natural gas for [3-5]:

1. Space heating by combustion in conventional furnaces
2. Space cooling by combustion in absorption-type air conditioners
3. Clothes-drying appliances
4. Stove-top and oven cooking
5. Hot water heating

## FIGURE 7.2

### U.S. Natural Gas Reserves

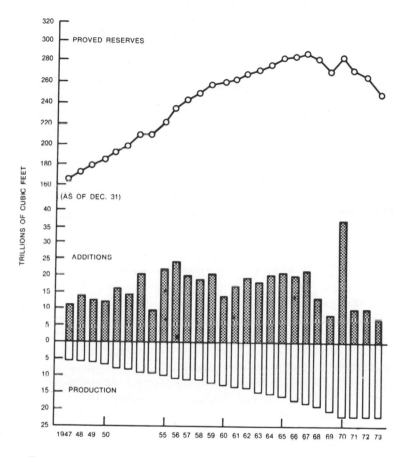

Source: Project independence report, reference 1.

Besides such conventional applications, hydrogen piped to residences or to commercial establishments offers the opportunity for some new applications, such as cooking and heating by means of flameless catalytic burners and generation of electricity with a fuel cell.

Flameless catalytic combustion of hydrogen is already a reasonably well-developed technology—temperatures can be varied between those low enough for mere warmth and high enough for cooking [3, 6]. Catalytic burners distributed through a building could be used for space relative difficulty of deploying leakproof pipe, compared with that of deploying electric cabling, brings the practicality of this concept into question. The fuel cell application suggests that if hydrogen were widely distributed to residences, an "all-hydrogen home" analogous to an "all-electric home" would be possible, although this would mean that the size of the fuel cell would have to be chosen to meet peak demands unless some kind of off-peak electric storage device, such as a battery, were used.

For safe use in residences, it is generally agreed that an "odorant" would have to be added to the odorless hydrogen so that people could smell leaks [3, 5, 6]. The addition of an odorant has long been required for residential use of methane (also odorless) in most states. It is also commonly accepted that for safety a colorant (illuminant) would have to be added to the hydrogen so that hydrogen's normally invisible flame could be seen [3, 5, 6].

Unfortunately, such odorants and colorants added to hydrogen gas makes the use of the catalytic burners and fuel cells more difficult. Operation of both these devices requires catalysts, which are readily "poisoned" (rendered less effective or useless) by the cheapest (sulfur-containing) odorant and colorant chemical compounds [3]. However, this drawback will probably be overcome by the development and use of alternative odorants and colorants or by additional research on and development of alternative catalysts.

The combustion of hydrogen requires burners designed for hydrogen. A simple change in fuel with no change in equipment gives unsatisfactory results. Flame propagation, temperature, and other physical combustion properties of hydrogen differ from methane, and smaller gas orifices are required in the burner for hydrogen than for methane. There is no technical barrier, however, to the production and use of hydrogen burners [3, 7]. Indeed, before natural gas was abundant, "town gas," "water gas," "coal gas," or "manufactured gas" (all mixtures of hydrogen, carbon monoxide, and other substances) were commonly piped to residences [3-6]. Thus, substantial experience concerning the design of burners for hydrogen-rich gases is available. [3].

Because the primary product of hydrogen combustion is (nontoxic) water vapor, it is often said that hydrogen offers an advantage over methane combustion for space heating because venting would not be necessary to exhaust toxic combustion products (especially carbon monoxide). Without venting, a structure could be sealed, thereby saving energy normally expended to heat newly infiltrated air. Moreover, in cold winter climates the buildup of humidity (if not carried to extremes) would increase human comfort in the normally dry, heated indoor air [3]. Unfortunately, this argument overlooks the need for air infiltration to provide the oxygen to support combustion—just as in methane systems. Many methane systems already avoid burning heated air by drawing combustion air directly from the outdoors. Consequently, the advantage of hydrogen for space heating is often overstated. Nevertheless, by eliminating the loss of heat up the flue, hydrogen should yield a net gain in efficiency.

Industrial use of hydrogen would go beyond the uses mentioned for residential and commercial applications to combustion for process heat and generation of process steam. These two uses consume about 30 percent and 45 percent, respectively, of all U.S. industrial energy use [8]. Currently, natural gas is highly favored by industry because of its artificially low cost and very clean combustion properties (which lower costs of pollution control). However, as the availability of natural gas diminishes, the industrial use of natural gas is being curtailed drastically.

Because of the clean combustion properties of hydrogen, industries would find special advantage in its combustion for process heat. Indeed, this clean combustion could warrant a price premium for hydrogen above alternative fuels because the expense of pollution control devices could be largely avoided (some control of $NO_x$ might still be needed). Again, use of the correct burner design would be essential.

The generation of process steam is probably the most attractive industrial use of hydrogen. When hydrogen is burned with pure oxygen the only combustion product is water vapor (steam) [3, 9]. As a result, it is feasible to generate high-quality, high-purity steam directly, without the use of conventional boilers [9]. Clearly, the elimination of boilers must be traded off against the added expense of providing pure oxygen either by purchase, by production in an air separation unit, or by saving and transporting the hydrogen coproduced with the hydrogen.

There are considerable advantages of producing steam in this manner. First, the steam is produced at a very high temperature— so high, in fact, that water must usually be injected to lower the temperatures so that conventional materials can handle the steam [9]. Second, the steam is very pure. Third, since not even $NO_x$ is produced and the sole combustion product is the steam, there is no need to provide controls for air pollutants. Unfortunately, however, if industry obtains its supply of hydrogen from the same source as the residential market, the required odorants and colorants will add a chemical impurity to the steam.*

The odorant and colorant safety additives also affect the potentially large industrial use of hydrogen as a chemical reactant. Chemi-

---

*It might be feasible, however, to inject the additives only at residential substations, thereby leaving the hydrogen uncontaminated for industry.

cal purification often would be needed to remove the safety additives before hydrogen could be used as a chemical. This impurity removal would not necessarily be a burdensome extra cost, however, because industries normally control the purity of incoming chemicals when they are to be used for high-purity applications. Today it is unclear what purity of hydrogen would be made available at the residential or industrial consumer level because the purity will largely depend on the means of hydrogen production and its storage and distribution en route to the consumer.

For many reasons, the gas industry prefers to gasify coal to produce methane rather than hydrogen because synthetic methane can be blended with natural gas supplied with no change being apparent to the customer and no change being required in gas utility local distribution networks. However, because even coal gasification would cease to be a cheap source of methane as increased demand for coal raised its price, a shift to nonfossil hydrogen might still be necessary later. That time is generally beyond the planning horizon of the troubled gas industry, however.

Chapter 16 further discusses gas utility applications.

### Electric Utilities

One of the most vexing problems of electric utilities is the variation in demand at different times of the day, different days of the week, and different months of the year [1-6, 10-16]. A sample load factor curve is shown in Figure 7.3. Yet, because for any given utility, variations in "load factor" are fairly predictable, utilities have generally found the means to cope with the variations. However, these variations in demand necessitate investment in generating equipment sized to meet peak demand. This means that much generating capacity is underutilized during slack demand periods while often strained to capacity at peak demand periods.

It is standard practice for utilities to categorize their load profiles into base, intermediate, and peak components [11, 12] as shown in Figure 7.3. Utilities allocate their most efficient and reliable generating equipment to base-load service. Consequently, this base-load equipment is typically the newest and most expensive in the utility. Because nuclear reactors are characterized by high investment costs (per unit output) and have difficulty varying their output to follow swings in demand, they are most productively used on a constant basis, operating at their full rated capacity. Currently, nuclear reactors and the best fossil-fuel-fired plants are always committed to base-load service [11, 12]. The older, less reliable, and less efficient fossil-fired plants are used for intermediate-load service, while peak-load power

FIGURE 7.3

Typical Load Curves for Electric Power Generation for One Week

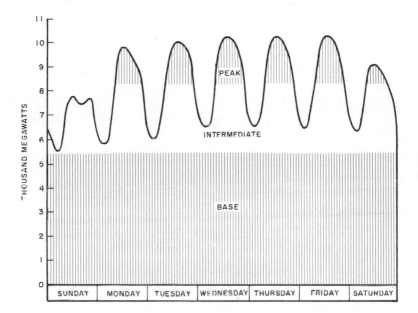

Note: Peak demand is often met by gas or petroleum fueled turbines.

is often supplied by inefficient turbine generators (similar to the engines on jet planes) that are costly to operate [12]. However, the high operating costs (for fuel) of turbines is somewhat offset by their low investment cost. When available, hydroelectric power is devoted to intermediate- or peak-load applications—except in regions with abundant hydroelectric power, where it can contribute to base load. To be able to meet very short-term variations of demand and as a hedge against equipment breakdown, utilities keep a small portion of their spare generating capacity in operation. The amount of this "spinning reserve" is often set in the form of a systems reliability criterion by the public utility commissions, which regulate utilities.

Utilities have shown great interest in finding new ways to follow the swings in the demand load factor, including institutional mechanisms (such as variable pricing) that would tend to smooth demand [17]. They have investigated several forms of energy storage, which in effect allows base-load facilities to meet some of the intermediate-load demand. This effect is shown in Figure 7.4. The most highly devel-

FIGURE 7.4

Average Winter Weekly Electric Power Load Curve

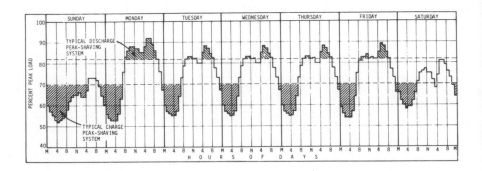

Source: Reference 12.

oped, although not widely used, form of energy storage is pumped hydroelectric storage [12]. This form of energy storage uses off-peak
power to pump water uphill to a reservoir; then, during the peak
demand periods, the water is released to flow back downhill and generate electricity [12]. However, since locations suitable for the large,
pumped hydroelectric storage facilities are not common, utilities
avidly seek technological analogues to pumped hydroelectric storage.

Hydrogen can be produced with off-peak power, stored, and then
later consumed to generate peak power, thereby serving as a chemical analogue to pumped hydroelectric storage [6, 10-12]. This approach is best matched to nuclear power because it provides a means
of converting the otherwise unstorable heat output of a reactor into a
storable fuel. There are several optional paths for the round trip between off-peak electricity through storage and generation back to
peak electricity; these options are shown in Figure 7.5. The round-
trip energy efficiency varies according to path, but for the liquid
hydrogen/fuel cell option it is only about 25 percent; for comparison,
pumped hydroelectric storage has a round-trip efficiency of about 66
percent [18]. Neither the hydrogen energy storage nor pumped hydroelectric storage is particularly efficient from an energy point of view,
but they offer utilities a means of attacking their peak-load problems.
As shown in Figure 7.5, the use of hydrogen as an energy storage mechanism paves the way for electric utilities to use fuel cells.

Public Service Electric and Gas (PSE&G) in New Jersey is now
experimenting—on a small scale—with the option shown graded in Fig-

FIGURE 7.5

Hydrogen Energy Storage Options for Electric Utilities

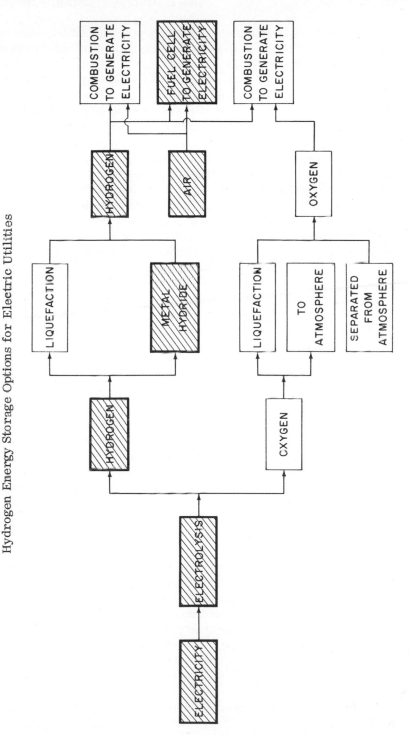

Note: Shading shows system under test at Public Service Electric & Gas in New Jersey.

81

ure 7.5; the iron-titanium metal hydride storage bed was developed
by Brookhaven National Laboratory and is now routinely operated at
PSE&G's laboratory [19].

Utilities are currently encountering difficulty in expanding their
facilities because of land use constraints and other environmental limitations. Nuclear power plant siting is proving especially difficult because of radiation hazard restrictions, cooling-water availability, and
limitations on the discharge of waste heat. As a result, nuclear power
plants are being sited farther and farther from the population centers
they serve, and, as a direct result, the amount of land needed for
power transmission corridors is directly increased. Aesthetic reasons have led to growing public opposition to deployment of new overhead high-voltage electric transmission lines and towers. In response
to the difficulties and delays in land acquisition for new corridors of
overhead power transmission, utilities are showing increased interest
in the high-cost technology of underground electric transmission for
areas of population density [20].

In the future, gaseous hydrogen sent in underground pipelines
may offer an aesthetically satisfactory and economically competitive
alternative to electric power for long-distance energy transmission
[3, 20]. It has been argued that remote siting of nuclear power plants
coupled with hydrogen production for energy transmission creates
several potentially attractive options to electric utilities, as shown in
Figure 7.6 [10-16, 21, 22]. Although the assumption has never been
tested, it is widely believed that, because of the much more favorable
aesthetic impact and much reduced commitment of land, it should prove
simpler for a utility to secure permission to deploy a single underground hydrogen pipeline* than to deploy an array of overhead electric transmission towers and power lines [23].

As discussed in Chapter 12, however, there is not a very good
economic match between the normal-sized nuclear power plant and
hydrogen pipelines. Use of hydrogen for energy transmission would
also facilitate the use of off-peak energy storage and the use of fuel
cells to generate electricity. Some utilities have suggested locating
the hydrogen-to-electricity generating facilities near the consumer
(at the substation level) through use of fuel cells run on hydrogen.
Because fuel cells are noiseless and fumeless, they should be good
neighbors [3]. Realization of this concept would require deployment
of hydrogen distribution pipelines in the city to serve the fuel-cell-
equipped substations. A major drawback to this concept of using hydrogen and fuel cells is the lowered net energy efficiency of delivered

_____

*For reasons of reliability, however, two pipelines might be de-
sirable.

FIGURE 7.6

Options for Using Hydrogen for Energy Transmission from Distant Nuclear Reactors

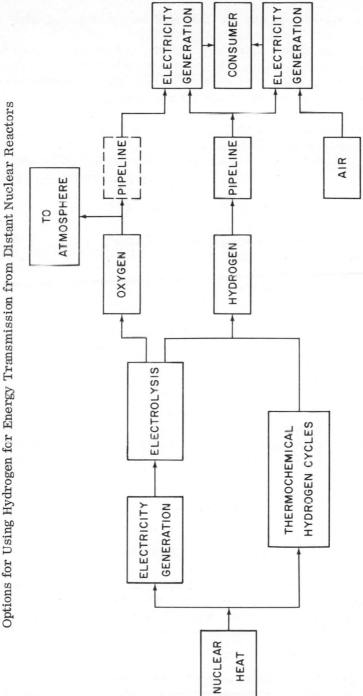

Note: The final electricity generation may be by combustion under a boiler, in a turbine, or in a fuel cell. Since transmission of the oxygen by pipeline is not required, it is shown dashed.

electric power (even when the losses of the alternative electric transmission are included in the comparison).

As long as the same electric utility that originally generated the
hydrogen consumed it itself to make electricity there would appear to
be few governmental regulatory barriers to the implementation of hydrogen pipelines as a substitute for high-voltage electric transmission.
Although some electric and gas utilities have suggested the sale of
some of their hydrogen [24], it is far from clear that public utilities
commissions would allow such sales.

Chapter 16 further discusses electric utility applications.

### Combined Electric and Gas Utilities

All the above advantages and disadvantages to the use of hydrogen apply to a single combined utility serving both gas and electric
markets.  There is, however, one additional and potentially very important advantage to a combined utility unavailable to separated utilities:  the ability to use hydrogen to load-level both the gas and the
electric systems [10].  Although they vary by region, climate, and
economic activity, the peak annual demands of most electric utilities
usually occur in the summer (because of electric air conditioners)
while gas utilities generally experience their peak demand in winter
(because of space heating).*  Thus, a combined utility that sold hydrogen rather than natural gas could load-level electricity in the summer
and gas in the winter with the same hydrogen generation and storage
facility.  The added flexibility and concomitant improvement in hydrogen facility utilization could be expected to have a favorable effect
upon the economics of hydrogen as an energy storage medium.

## AUTOMOTIVE APPLICATIONS

### Private Passenger Vehicles

There is no doubt at all that hydrogen can be used in internal
combustion engines to power automobiles [25-30].  Many well-pub-
licized demonstrations have shown that with relatively minor and
simple adaptations engines run well and cleanly on hydrogen and air

-------------------------

*Electric utilities advertised to promote electric space heating
a few years ago in an attempt to create winter demand equivalent to
summer demand.

(even more cleanly on pure hydrogen and oxygen because $NO_x$ cannot form). The ease of adapting stock engines has brought much attention to the clean air advantages of hydrogen fuel in place of gasoline or other chemicals containing carbon [25-34]. In addition, some claims have been made that hydrogen increases the efficiency of internal combustion engines [35, 36].* In an environmentally conscious and energy-short world, both decreased emissions of air pollutants and increased engine efficiency are attractive attributes. The clean air advantage of hydrogen should remain even if external combustion engines displace the internal combustion engine.

There are some notable offsetting disadvantages to the use of hydrogen in automobiles. There are really only three potentially practical methods to distribute and store hydrogen for automotive use: as a liquid, in the form of metal hydrides, or in the form of hydrogen-containing chemical compounds. The last option would include ammonia, $NH_3$, methanol, $CH_3OH$, and others. But, as noted earlier, these compounds have been excluded from our definition of the hydrogen energy economy. The remaining two options are both awkward and fraught with difficulties compared with other fuel options.

Use of liquid hydrogen would require the use of cryogenic storage vessels, which are likely to remain bulky and costly compared with the simple sheet metal tank commonly used to hold gasoline—and which would also be suitable for alcohols like methanol. In the small sizes (about 60 gallons) appropriate for hydrogen storage in automobiles, the cryogenic vessels would be thermally inefficient, with a boil-off rate of about 1 to 2 percent per day [37]. However, if the car were used regularly, thereby consuming the boil-off hydrogen, no fuel would actually be lost. By far the greatest barrier to the use of liquid hydrogen in automobiles would be the establishment of a dense liquid hydrogen distribution network.

The energy efficiency of the total system of hydrogen production liquefaction, distribution, storage, transfer, and consumption in automobiles is a parameter of importance. Figure 7.7 shows several options for this chain and compares it to synthetic gasoline from coal. As noted in Chapter 5, the liquefaction of hydrogen is only about 77 percent efficient in net energy terms [38]. During transfer of liquid hydrogen from one cryogenic storage vessel to another, there is unavoidable boil-off. For reasons of safety (and economics), there can

---

*It should be noted, however, that large-sized engines generally show better thermal efficiency (because of a smaller surface to volume ratio in the combustion chamber) than small engines. Thus, the trend toward smaller, lighter cars with smaller engines now begun in order to decrease automotive fuel consumption is likely to lessen the net effect of the improved efficiency reported for hydrogen.

be little doubt that this boil-off would be captured and possibly recycled [33, 39], but even reliquefaction of the already cold gas would
require a considerable expenditure of energy—especially if small liquefaction units were used. The energy losses associated with unavoidable boil-off and reliquefaction at the several transfer points
would tend to offset gains in engine efficiency.

Metal hydrides seem to offer an attractive alternative to liquid
hydrogen for automotive use [40, 41]. Many of the candidate hydrides,
however, can be eliminated because they do not possess physical properties well suited to the rigors of automotive use. First, many are
unable to release the hydrogen fast enough to keep the car operating
unless they are held at a high temperature (around 600°F). Second, the
widespread belief that heat from the automotive exhaust system would
be adequate to release the needed hydrogen is apparently incorrect—
real-world heat exchangers are not efficient enough to strike this balance for otherwise suitable metal hydrides [42, 43]. Third, and perhaps most detrimental to the case for metal hydrides, is the heavy
weight and large size (about 700 pounds and 11 cubic feet for even the
lightest candidate hydride, $MgH_2$) needed to contain enough hydrogen
to provide a cruising range equivalent to that provided by present automobile gas tanks in full-sized American cars [44]. The present
trend toward smaller, lighter automobiles helps the case for metal
hydrides on the one hand because the decreased fuel consumption would
lessen the weight of the hydrides needed; but, on the other hand, automobile designers would probably be reluctant to relinquish the precious
space and weight allocations in a small car to a heavy metal hydride
bed.

Essentially three options are available for distributing and refueling metal hydride beds in automobiles: gas recharge in a filling
station, gas recharge in residences, and physical exchange of hydride
beds at a filling station. The first two approaches require either the
prior existence of a gaseous hydrogen distribution system or small-
scale electrolysis units. The heat transfer problems of metal hydride
recharge suggest that refueling in a filling station might take longer
( 15 to 30 minutes [44]) than consumers would accept. Recharging at
home at night might prove acceptable provided the user never strayed
far from home.

The physical exchange of an entire depleted metal hydride bed
at filling stations loses its intuitive appeal when the awkwardness of
accomplishing the rapid exchange of about 700 pounds of bulky material is considered. Moreover, in a viable exchange system the variation allowed in the shape and size of hydride beds would have to be
severely limited. Also, if gaseous hydrogen were piped to the filling
station, the beds could be recharged there, but if a gaseous hydrogen
distribution system did not already exist, added burden and energy in-

# FIGURE 7.7

## Automotive Synthetic Fuels Systems Comparison

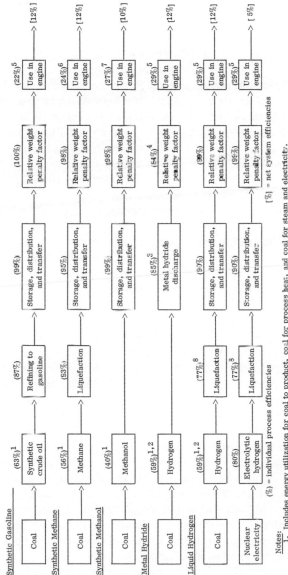

Notes:
1. Includes energy utilization for coal to product, coal for process heat, and coal for steam and electricity.
2. SRI internal data.
3. Assumes that although most of the heat needed to discharge the hydride comes from the heat in the exhaust and the cooling water, 15 percent extra heat is required for self-sustaining operation (see reference 36).
4. Assumes 530 lb metal hydride plus tank; this weight includes a correction to account for the effect on vehicle range resulting from the superior thermal efficiency of hydrogen in engines.
5. Reference 36.
6. J. Pangborn et al., "Alternative Fuels for Automotive Transportation—A Feasibility Study" (U.S. Environmental Protection Agency), 1974.
7. H. Heitland et al., "Comparative Results on Methanol and Gasoline Fueled Passenger Cars" (Second Symposium on Low Pollution Power Systems Development, Dusseldorf, Germany), November 1974.
8. Reference 38.

efficiency of shipping the hydride beds to and from a central recharging point would be entailed.

The attractiveness of hydrogen for private automobiles purely on the basis of engine compatibility and the cause of clean air seems to be more than offset by the new fuel distribution networks and refueling procedures required.

## Fleet Vehicles

Fleets of cars, trucks, and buses presently consume about 30 percent of all fuel used in the automotive sector. Because trucks and buses are bulky themselves, the bulkiness of a large liquid hydrogen or metal hydride fuel system would be less constraining than in automobiles.

Many fleet vehicles often do not travel far from the home terminals where they are refueled. Moreover, because they frequently are idle for much of the night, a prolonged refueling operation—as for metal hydride storage—need not be as constraining. Many other fleet vehicles operate repetitively between fixed end-points. Thus, the complications of refueling with either liquid hydrogen or a slow recharging of a metal hydride could often be tolerated for fleet vehicles. Consequently, the use of hydrogen could be implemented far more readily in fleet vehicles than in private automobiles.

## Off-the-Road Vehicles

While, in principle, off-the-road vehicles such as earthmovers, farm machinery, mining machinery, snowmobiles, motorcycles, forklifts, and so forth could all be run on hydrogen, the problems of on-board fuel storage and distribution to the vehicles is acute. There are a few exceptional cases, however, in which the especially advantageous properties of hydrogen combustion may offset these difficulties. Forklifts used indoors in warehouses are one of the prime examples [45]. Already, to avoid carbon monoxide poisoning in confined environments, most forklifts used indoors are either battery-electric or operated on relatively clean-burning liquid propane, butane, or compressed methane. Hydrogen combustion's complete lack of toxic carbon monoxide exhaust products would, therefore, place a hydrogen-fueled forklift on a par with an electric forklift—as far as emissions are concerned. In such uses, the occupational health advantages of hydrogen could even justify payment of a premium price for the fuel.

Chapter 14 further discusses automotive applications.

## AIRCRAFT APPLICATIONS

Tests have proved that jet aircraft engines run well and very cleanly on hydrogen [45-48]. Under contract to NASA, Lockheed has recently completed a thoughtful analysis and evaluation of the design options available for hydrogen-fueled supersonic [49] and subsonic [50] transport aircraft. For both kinds of aircraft, the only viable form of hydrogen storage is the cryogenic liquid.

For hypersonic (more than five times the speed of sound) aircraft, liquid hydrogen is vastly superior to conventional jet fuels because it can be used en route to the engines to cool the aircraft skin, which is heated by aerodynamic drag. This advantageous synergism allows the use of lightweight materials such as aluminum instead of heavy, high-temperature materials, thereby reducing the weight of the aircraft [51-54].

Hydrogen's advantage over conventional jet fuel is not as great in subsonic aircraft as in hypersonic aircraft. The recent studies by Lockheed have shown that an optimized liquid-hydrogen-fueled subsonic passenger plane would possess the following attributes (compared with an advanced design airplane using conventional jet fuel) [50]:

1. Less gross takeoff weight
2. Nearly equal empty weight
3. Lower emissions of air pollutants (only nitrogen oxides)
4. Shorter takeoff distance (about 36 percent reduction)
5. Less noise in the takeoff zone
6. Slightly more noise in the landing zone
7. Slightly higher aircraft sales price (about 3 percent)
8. Lower energy utilization per passenger mile

These and other comparisons are shown in Table 7.1 for both transoceanic and transcontinental versions. It must be emphasized that in the Lockheed study it was assumed that future airplanes using either fuel will be less noisy than the quietest U.S.-made airplanes now flying [50].*

Because liquid hydrogen contains only about one-third the energy in the same volume as the conventional fuel (Jet-A, essentially kerosene), the volume required for fuel storage is much larger in a hydrogen-fueled airplane than in a conventional jet. Although today's commercial airplanes carry Jet-A fuel in tanks formed by the wing structure, hydrogen cannot be carried in the wings because of insufficient

---

*The wide-body Lockheed L1011, McDonnell-Douglas DC-10, and Boeing 747.

TABLE 7.1

Comparison of LH$_2$ and Jet-A Passenger Aircraft

| | Unit | 5,500 nmi, Mach 0.85 | | 3,000 nmi, Mach 0.85 | |
|---|---|---|---|---|---|
| | | LH$_2$ | Jet-A | LH$_2$ | Jet-A |
| Gross weight | lb | 391,700 | 523,200 | 335,200 | 404,400 |
| Operating empty weight | lb | 242,100 | 244,400 | 212,900 | 210,600 |
| Block fuel weight | lb | 52,900 | 165,500 | 28,000 | 82,300 |
| Thrust per engine | lb | 28,700 | 32,700 | 24,720 | 25,770 |
| Span | ft | 174 | 194.1 | 165.6 | 170.6 |
| Height | ft | 59.5 | 60.2 | 55.4 | 56.0 |
| Fuselage length | ft | 219 | 197 | 210 | 197 |
| Wing area | ft$^2$ | 3,363 | 4,186 | 3,047 | 3,235 |
| Takeoff distance | ft | 6,240 | 7,990 | 5,860 | 7,980 |
| Landing distance | ft | 5,810 | 5,210 | 5,804 | 5,760 |
| Aircraft price | millions of dollars | 26.9 | 26.5 | 23.4 | 22.6 |
| Energy utilization | Btu per seat nmi | 1,239 | 1,384 | 1,204 | 1,260 |

nmi = nautical mile

Source: Lockheed, reference 50.

space and also because the shape of the available space would result
in very inefficient containment (high surface to volume ratio).  Lockheed considered several locations for liquid hydrogen fuel storage,
including large over-the-wing pods; tanks inside the fuselage, either
above or below the passenger cabin; and tanks inside the fuselage in
front of and behind the passenger cabin [50].  Lockheed concluded
that from safety, structural, and effectiveness points of view the last
option was superior [50].  Figure 7.8 (a and b) shows the designs
Lockheed considered; the internal tank version was selected for the
most detailed analysis.  As shown in Figures 7.8b and 7.9, the selected airplane looks remarkably like conventional wide-body jets from
both the side and plan views.  The large size of the fuel storage volume makes it fairly clear that large, long-range aircraft are best
suited for liquid hydrogen fuel.

A commercial fleet of liquid hydrogen airplanes would require
the production of vast quantities of hydrogen and a large throughput
in the logistics system.  Such a fuel network, however, would require
relatively few distribution points [55] because the top 25 airports in
the country handle about 76 percent of all the air passengers, and the
top 10 airports handle about 70 percent of all the passengers [56].
Thus, with as few as 10 airports equipped to dispense the fuel, a
great proportion of the long-distance travel in liquid-hydrogen-fueled
aircraft could be accommodated.

Small general aviation aircraft could, in principle, use liquid
hydrogen, but the large size of the fuel tanks is an impediment.  In
addition, it would be extremely difficult to supply liquid hydrogen to
the large number of general aviation airports and landing strips.


## SHIP, TRAIN, AND SPACECRAFT APPLICATIONS

Hydrogen's demonstrated suitability for combustion in nearly
every class of engine implies that there should be no major technological barriers to its use in either ships or trains.  Liquid hydrogen is
already an essential fuel for spacecraft, and this single use constitutes
by far the largest contemporary market for merchant hydrogen (as opposed to captively produced and used hydrogen).

Since ships are not as limited by their size and shape as automobiles and trucks, the extra volume needed to store liquid hydrogen
should not pose much of a problem.  Storage as a metal hydride, however, probably involves too much weight to be taken seriously.  Yet
there does not seem to be much advantage to or incentive for the use of
hydrogen in large ships.  Ships are generally fueled either by a low-
grade heavy oil (bunker fuel) or by diesel.  Air pollution caused by
ships—especially oceangoing ships—has not been a problem because

# FIGURE 7.8

## Options for Hydrogen Storage in Aircraft

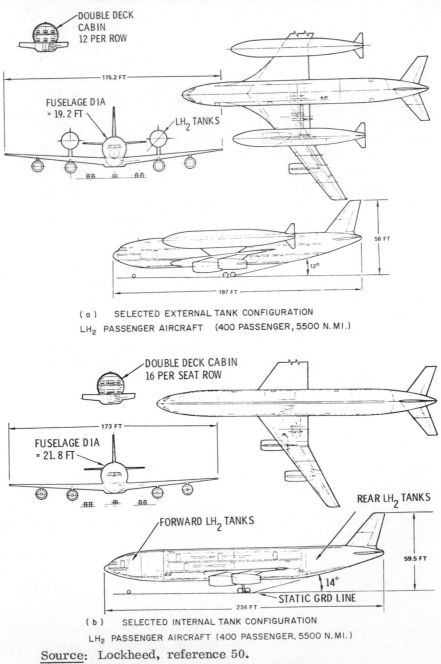

( a )    SELECTED EXTERNAL TANK CONFIGURATION
LH$_2$ PASSENGER AIRCRAFT   (400 PASSENGER, 5500 N.MI.)

( b )    SELECTED INTERNAL TANK CONFIGURATION
LH$_2$ PASSENGER AIRCRAFT (400 PASSENGER, 5500 N.MI.)

Source: Lockheed, reference 50.

FIGURE 7.9

Size Comparison:  LH$_2$ Versus Jet-A Passenger Aircraft

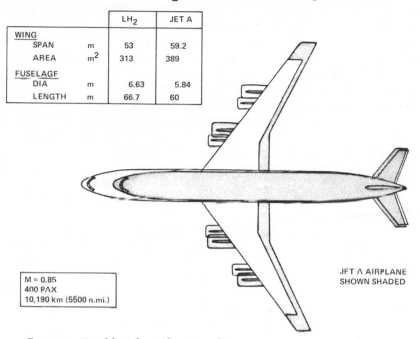

|  |  | LH$_2$ | JET A |
|---|---|---|---|
| **WING** |  |  |  |
| SPAN | m | 53 | 59.2 |
| AREA | m$^2$ | 313 | 389 |
| **FUSELAGE** |  |  |  |
| DIA | m | 6.63 | 5.84 |
| LENGTH | m | 66.7 | 60 |

JET A AIRPLANE
SHOWN SHADED

M = 0.85
400 PAX
10,190 km (5500 n.mi.)

Source:  Lockheed, reference 50.

most of their operations are far from population centers and, more
important, beyond the regulatory control of any government.  The
need to establish a global liquid hydrogen network would be a strong
disincentive to the conversion of oceangoing ships away from fuel oil.

Trains are not strongly constrained in their total volume, although they are severely limited in cross-section (set by the clearance
dimensions of tunnels, bridges, parallel tracks, and so on).  Use of
hydrogen to fuel gas turbine trains would certainly reduce air pollution
and, possibly, engine noise.  Advanced-design turbine-powered trains
are under development both in this country and abroad, although the
effort receives only modest funding.  Getting the fuel to the trains
would be relatively simple compared to other land transportation applications of hydrogen.  The train can carry a large fuel supply and
can be operated between fixed, though distant, end-points, thereby
greatly reducing the number of fueling points needed.

It has now been demonstrated (with models) that magnetic levitation using superconducting electromagnets is technically feasible for
a train running on special guideways [57, 58].  To date, all work on

this problem has assumed that liquid helium would be used as the cryogenic coolant for the magnets because superconductors with transition temperatures* higher than the normal boiling point of liquid hydrogen have been found only recently. The expectation that superconductors with still higher transition temperatures may yet be found means that in the future liquid hydrogen might serve as the cryogenic coolant instead of the more expensive and more scarce liquid helium. Conceivably, a magnetic levitation train could utilize the hydrogen both as a coolant and then as fuel. Such application is many years away and would probably be confined to high-speed passenger trains.

For spacecraft, liquid hydrogen is a nearly ideal fuel because its very high gravimetric energy density greatly reduces the total launch weight of a rocket. Liquid hydrogen (combusted with pure oxygen) is the standard fuel for the second and third stages of U.S. space exploration rockets. However, because of the need to maintain constant readiness, hydrogen is not now used extensively in present nuclear warhead missiles.

NASA is still developing and designing the space shuttle, a reusable rocket with a high payload capability that is envisoned for use in space activities close to earth. For example, the proposed large satellite-borne solar energy collector that would transmit the energy to earth on a microwave beam would use the space shuttle to ferry construction material into orbit. Construction of this huge solar energy device would require from 300 to 1,000 ferry trips of a second-generation space shuttle [59]. Although deployment of this solar energy satellite is questionable, this or comparable use of the space shuttle would require vast quantities of hydrogen.

Chapter 15 further discusses aviation applications.


MILITARY APPLICATIONS

Just as in the civilian sector, there is potential for military use of hydrogen in cars, trucks, buses, ships, and airplanes. There has been discussion (mainly among civilians) of a nuclear-powered aircraft carrier that could use electrolytically derived liquid hydrogen to fuel the carriers' jets.

---

*Superconductors are characterized by several physical parameters. The "transition temperature" is the temperature at which the material undergoes a transition from a normal, resistive conductor to a superconducting, resistanceless material. To operate superconducting equipment at the normal liquid hydrogen boiling temperatures of $20.4°K$ requires a superconductor with a transition temperature several degrees higher.

At first sight, the ability to make hydrogen anywhere in the world from any basic energy resource would seem to have considerable appeal to the military because it might eliminate burdensome, long-distance energy supply logistics systems.

With more thought, however, the attractiveness of hydrogen in military applications fades for several reasons. First, the bulkiness of hydrogen fuel storage contradicts the need for small, sleek, highly maneuverable, supersonic fighter aircraft; bulkiness also discourages hydrogen use in land vehicles such as tanks because it would increase their size as a target. Second, hydrogen production could not completely eliminate the logistics supply line because in most cases some basic energy resource would still have to be delivered. Third, the extra procedures and precautions needed to liquefy and handle vast quantities of cryogenic hydrogen are not very compatible with combat. Fourth, the military requires great operational flexibility, and it would be undesirable to have some airplanes, for example, fueled with hydrogen while others used conventional jet fuel.

Military use of hydrogen seems to compromise, rather than improve, readiness and flexibility, and to increase vulnerability. This no doubt accounts for the present low level of enthusiasm in both the civilian and uniformed defense establishment for military use of hydrogen as a fuel.

This lack of military interest has very important ramifications for the evolution of the hydrogen energy economy and the need for civilian-sponsored research and development. Historically, military needs have been the stimulus for much technological development with subsequent "spin-off" into the civilian sector.

## CHEMICAL APPLICATIONS

Besides use as a fuel, hydrogen has many applications as a chemical because it is an excellent reducing agent.* It is essential in reforming hydrocarbons (such as making plastics from oil), and it is necessary for synthesizing other important chemicals such as ammonia. It would be naive to expect that the hydrogen energy economy concept could develop without strong synergisms developing between it and the hydrogen chemical economy [4].

---

*Chemical reduction is essentially the inverse of oxidation, and in many reducing applications, hydrogen is used to remove oxygen from a compound. For example, iron oxide, $Fe_2O_3$, can be reduced with hydrogen to yield iron.

## Ammonia Synthesis

Ammonia, $NH_3$, is one of the most basic chemicals used in modern industrial society [60]. It is used as a feedstock for many chemical processes and as a fertilizer*—either directly or transformed into ammonium sulfate, nitrate, or urea [60]. Recently, production of ammonia has grown rapidly, and the increasing emphasis on expanded agricultural output by putting more land in cultivation is likely to spur yet more growth. However, growth in the use of ammonia as a fertilizer is bound to level off soon in the United States because fertilizer application rates are nearing their optimum economic usage [61, 62].

At the present time, production of essentially all ammonia in the United States is accomplished by synthesis from a nitrogen–hydrogen gas mixture derived from air and methane [4, 61, 63]. Methane is reformed (by the process described in Chapter 4) to yield hydrogen. The required nitrogen is obtained by burning methane in air to consume the oxygen in the air and then chemically separating the carbon dioxide and water combustion products to yield the nitrogen [61].† The heat derived from the methane combustion is used as process heat.

Ammonia synthesis from a methane feedstock is a very well-understood, highly developed, highly efficient technology in which there is little room for improvement [61]. In the United States, methane has traditionally been the source of both hydrogen and process heat because it has been abundant and cheap [61]. Most ammonia plants have been located near the methane sources; the majority are now located in an arc around the U.S. Gulf of Mexico coast and the Southwest [61].

It is frequently stated that ammonia synthesis will provide a large market for hydrogen in the future hydrogen economy [3-6]. The reasoning has been that as methane supplies dwindle and the price rises, ammonia producers will be priced out of methane. They would then turn to hydrogen as a feedstock provided it were available for purchase; alternatively, they might make it themselves electrolytically. This line of reasoning has been fostered by a misunderstanding of the design of actual ammonia synthesis plants. In particular, because the role of methane in providing the nitrogen is not widely known, it was easy to assume that the abandonment of methane as a feedstock would only affect the plant's source of hydrogen. But, in fact, because

---

*Nitrogen is a key element in controlling plant growth. The nitrogen in ammonia or the ammonium ion is readily available for plant uptake.

†Air is composed mainly of gaseous oxygen (about 20 percent) and gaseous nitrogen (about 80 percent).

the designs have been so highly integrated, abandonment of methane would necessitate redesign of nearly the entire plant [61]. Experts in ammonia production report that if methane were not available as a feedstock to an ammonia synthesis plant, it would prove more effective to turn to nearly any alternative liquid or gaseous hydrocarbon in preference to hydrogen, because this would result in the smallest overall change in existing plant designs and operations [61]. However, if hydrogen became available at a lower cost than it could be obtained from methane or alternative hydrocarbon fuels, it is probable that new ammonia plants would be designed to use the lower-cost hydrogen feedstock [61]. There is no inherent technological reason why new ammonia plants could not be designed to use a hydrogen feedstock. Chapter 17 further discusses the likelihood of ammonia synthesis becoming a significant portion of the hydrogen economy.

## Coal Gasification or Liquefaction

As noted in Chapters 4 and 11, for the foreseeable future the cheapest source of hydrogen not based on liquid or gaseous fossil fuels will be coal gasification. Yet coal gasification (and also liquefaction) processes that yield methane (or synthetic crude oil) could be made more efficient in their use of coal if an independent source of hydrogen were available. Some, therefore, envision hydrogen derived from nuclear energy finding a market in the coal gasification or liquefaction industry [5].

This proposed application of merchant hydrogen, however, is not very likely to be realized. In the first place, it would prove less efficient to use nuclear power to produce hydrogen as a distinct intermediate product than to use the nuclear power for process heat in the gasification or liquefaction process. Second, if nuclear-derived hydrogen could be purchased for the process more cheaply than it could be produced from the coal, it would make little economic sense to manufacture high-cost synthetic methane when low-cost hydrogen was available as a competitor. Thus, it appears that although coal gasification and liquefaction technologies will eventually be deployed and will require hydrogen, the hydrogen will be derived from the energy of coal itself. Similar conclusions apply to the hydrogenation steps in oil shale processing and the production of methanol from coal and various waste materials.

## Minor Chemical Uses

Hydrogen is used today in the food industry to transform unsaturated fats to saturated fats. This occurs, for example, in the mar-

garine, peanut butter, and shortening industries [4]. These chemical
feedstock uses of hydrogen would constitute only a tiny portion of the
hydrogen economy. Moreover, they do not appear destined to play a
pivotal role in the hydrogen economy evolution and are not considered
further.

## Chemical Reduction

Many industries employ chemical reduction, for example, to
transform an oxide or sulfide metal ore into a raw metal, and to remove oxidation from a surface prior to plating or finishing. Steelmaking with hydrogen used as an iron ore reducing agent is becoming
increasingly attractive and is likely to develop into a major use of hydrogen [4, 64–66].

In steelmaking, the ore is iron oxide. The first step in steelmaking is to chemically reduce the ore to raw iron. To achieve this
reduction, a special class of coal, called "metallurgical" or "coking"
coal is required. Coal is turned into coke, a spongy form of pure
carbon [65], by a destructive distillation process. Volatile gases
driven off during the coking process are burned to provide the process
heat [65]. Coke and raw iron ore are then mixed together and heated
[65]. Because the oxygen is more attracted to the carbon than to the
iron at elevated temperatures, the ore is chemically changed into
(impure) metallic iron and the carbon combines with the oxygen to
yield carbon dioxide ($CO_2$) and some carbon monoxide (CO). In the
early days of steelmaking, charcoal was used instead of coke. The
combined need for both iron ore and coal dictated the locations of iron
and steelmaking installations [65].

Most large steel companies in the United States are vertically
integrated and control and mine their own resources of coking coal.
Not all countries are as well endowed with coal suitable for coking as
is the United States, and considerable quantities of eastern coal are
exported each year for steelmaking—especially to Japan, which has
no suitable indigenous resources [64]. Coking coal resources are becoming increasingly scarce and expensive [65], and the control of
air pollution from coke making and ore reduction is increasingly being forced on a steel industry reluctant to make the investments in the
necessary air pollution control equipment [67–69]. Both factors are
enhancing interest in new ways of steelmaking

Nuclear steelmaking and hydrogen used for ore reduction are
being looked on with new favor [64, 66]. As presently conceived,
nuclear steelmaking uses the heat of a nuclear reactor both to yield
process heat and to generate electricity for the electric arc furnaces,
which melt the raw iron for transformation into steel [64]. Part of

the nuclear energy could be used to make hydrogen for ore reduction. The source of hydrogen envisioned is usually methane, with the nuclear reactor supplying the heat needed for reformation [64, 66]. However, as methane becomes less available, no doubt emphasis will shift to obtaining the hydrogen either by electrolysis or by thermochemical cycles. In any event, the vast quantities of hydrogen involved and the presence of an on-site nuclear reactor suggest that this hydrogen will be produced captively rather than purchased.

Currently the major interest in nuclear steelmaking and hydrogen reduction of ore is found in foreign nations—especially Japan and England [64]. However, in this country, Bethlehem Steel is apparently taking some interest [66].

Chapter 17 further discusses the use of hydrogen in steelmaking.

## CRYOGENIC APPLICATIONS

Liquid hydrogen has the second lowest boiling point of any liquid, as can be seen in Table 7.2. Because until recently no materials had been found that were superconducting above 20.4°K, there was no possibility of using liquid hydrogen to cool superconducting equipment. Recently, however, it has been found that $Nb_3Ge$ superconducts at about 23°K and below [70-72]. Thus, there is widespread hope that other materials will be found with even higher superconducting transition temperatures, thereby opening the way to use of liquid hydrogen as a coolant instead of the usual liquid helium [71].

This recent discovery is very important for several reasons. First, it takes more energy to liquefy helium than hydrogen. (The theoretical minimum for helium is about twice that of hydrogen [73].) Second, helium is a rare element mainly found in a dilute (2 percent) association with natural gas from a few fields [74, 75]. Although the United States has most of the world's reserves of helium, and until recently was building a stockpile for future use, additions to the storage program have been terminated by the federal government [75]. Consequently, helium is now being wasted into the atmosphere as the natural gas is burned. Moreover, only two gases, helium and hydrogen, are light enough to escape the earth's gravity; hence, helium diffuses through the atmosphere and is permanently lost into outer space. Many technologists question whether terrestrial reserves of helium will be adequate to sustain large-scale industrial use of superconducting technology unless the wastage of helium is stopped [74, 75].

If additional and useful new materials are found that will superconduct in liquid hydrogen, important new uses [75] for the hydrogen will be established. Important synergisms with other aspects of the hydrogen economy would be inevitable.

TABLE 7.2

Normal Boiling Points of Cryogenic Liquids

| Element | Temperature | | |
|---------|------|------|------|
|  | (°K) | (°C) | (°F) |
| Helium (He) | 4.2 | −269 | −452 |
| Hydrogen ($H_2$) | 20.4 | −253 | −423 |
| Neon (Ne) | 27 | −246 | −411 |
| Nitrogen ($N_2$) | 77 | −196 | −320 |
| Argon (Ar) | 87 | −186 | −303 |
| Oxygen ($O_2$) | 90 | −183 | −297 |
| Methane ($CH_4$) | 112 | −162 | −259 |

Uses of cryogenic liquids not involving superconductors are increasing. Today, nearly all these applications use either liquid helium or the far cheaper liquid nitrogen. Once widely available commercially, liquid hydrogen, with its low temperature, would surely find many applications in a refrigeration/fuel synergism. Some of the diverse potential uses of cryogenic liquids are refrigeration, freeze drying of foods [77], and embrittlement of materials to enhance fracturing as a prelude to separation and recycling [78, 79]. It is likely that boiloff from liquid hydrogen used as a coolant would be trapped and then consumed as a fuel by the user.

REFERENCES

1. U.S., Federal Energy Administration, Project Independence (Washington, D.C., November 1974).

2. G. T. Kinney, "U.S. Gas Picture Darkens While SNG and LNG Lag," Oil and Gas Journal, June 16, 1975, pp. 17-20.

3. D. P. Gregory et al., "A Hydrogen-Energy System," American Gas Association, Alexandria, Va., August 1972.

4. E. Fein, "A Hydrogen Based Energy Economy," The Futures Group, Glastonbury, Conn., October 1972.

5. L. T. Blank et al., "A Hydrogen Energy Carrier," Systems Design Institute, National Aeronautics and Space Administration-American Society for Engineering Education, Johnson Space Center, Houston, Texas, 1973.

6. "Hydrogen and Other Synthetic Fuels," a summary of the work of the Synthetic Fuels Panel, prepared for the Federal Council on Science and Technology R&D Goals Study, September 1972.

7. N. R. Baker, "Oxides of Nitrogen Control Techniques for Appliance Conversion to Hydrogen Fuel," Ninth Intersociety Energy Conversion Engineering Conference, 1974, pp. 463-67.

8. "Patterns of Energy Consumption in the United States," Stanford Research Institute for the Office of Science and Technology, Washington, D.C., January 1972.

9. W. Hausz, "Hydrogen and the Energy Market," in Proceedings of the Cornell International Symposium and Workshop on the Hydrogen Economy, ed. S. Linke (Ithaca, N.Y.: Cornell University, April 1975), pp. 217-34.

10. P. A. Lewis, "Hydrogen Use by Energy Utilities," in Proceedings of the Cornell International Symposium and Workshop on the Hydrogen Economy, ed. S. Linke (Ithaca, N.Y.: Cornell University, April 1975), pp. 266-73.

11. P. A. Lewis and J. Zemkoski, "Prospects for Applying Electrochemical Energy Storage in Future Electric Power Systems," IEEE Intercon, Tec. Paper (paper delivered at IEEE meeting, New York, March 26-30).

12. R. A. Fernandes, "Hydrogen Cycle Peak-Sharing for Electric Utilities," Ninth Intersociety Energy Conversion Engineering Conference, 1974, pp. 413-22.

13. M. Lotker, E. Fein, and F. Salzano, "The Hydrogen Economy—A Utility Perspective," paper presented at the IEEE winter meeting, 1973.

14. J. M. Burger, "An Energy Utility Company's View of Hydrogen Energy," in Hydrogen Energy Fundamentals: A Symposium Course, ed. T. N. Veziroglu, presented at University of Miami, Miami Beach, Fla., March 3-5, 1975.

15. "U.S. Energy Prospects: An Engineering Viewpoint," a report by the Task Force on Energy of the National Academy of Engineering, Washington, D.C., 1974.

16. W. E. Winsche et al., "Hydrogen: Its Future Role in the Nation's Energy Economy," Science 180 (June 29, 1973): 1325-32.

17. "Utilities: Weak Point in the Energy Future," Business Week, January 20, 1975, pp. 46-54.

18. A. L. Robinson, "Energy Storage (I): Using Electricity More Efficiently," Science 184, no. 4139 (May 17, 1974): 785-87.

19. J. Burger et al., "Energy Storage for Utilities via Hydrogen Systems," Ninth Intersociety Energy Conversion Engineering Conference, 1974, pp. 428-34.

20. Raymond Huse, Public Service Electric and Gas, Newark, N.J., personal communication, 1974.

21. "Fuel Cell Research Finally Paying Off," Chemical and Engineering News, January 7, 1974, pp. 31-32.

22. T. H. Maugh II, "Fuel Cells: Dispersed Generation of Electricity," Science 178 (December 22, 1972): 1273-74B.

23.  D. P. Gregory, "A New Concept in Energy Transmission," Public Utilities Fortnightly, February 3, 1972, pp. 3-11.

24.  M. Lotker, "Hydrogen for the Electric Utilities—Long Range Possibilities," Ninth Intersociety Energy Conversion Engineering Conference, 1974, pp. 423-27.

25.  R. E. Billings and F. E. Lynch, "Performance and Nitric Oxide Control Parameters of the Hydrogen Engine," publication No. 73002, Billings Energy Research Corporation, Provo, Utah, April 1973.

26.  J. G. Finegold and W. D. Van Vorst, "Engine Performance with Gasoline and Hydrogen: A Comparative Study," in Hydrogen Energy, ed. T. N. Veziroglu (New York: Plenum, 1975), pp. 685-96.

27.  R. J. Schoeppel, "Design Criteria for Hydrogen Burning Engines," final report, EPA Contract EHS-70-103, October 1971.

28.  R. R. Adt et al., "The Hydrogen-Air Fueled Automobile Engine," Eighth Intersociety Energy Conversion Engineering Conference, 1973, pp. 194-97.

29.  K. H. Weil, "The Hydrogen I. C. Engine—Its Origins and Future in the Emerging Energy-Transportation-Environmental System," Seventh Intersociety Energy Conversion Engineering Conference, 1972, pp. 1355-63.

30.  R. G. Murray et al., "The Hydrogen Engine in Perspective," Seventh Intersociety Energy Conversion Engineering Conference, 1972, pp. 1375-81.

31.  W. J. D. Escher, "On the Higher Energy Form of Water ($H_2O^*$) in Automotive Vehicle Advanced Power Systems," Seventh Intersociety Energy Conversion Engineering Conference, 1972, p. 1392-1402.

32.  A. L. Austin, "A Survey of Hydrogen's Potential as a Vehicular Fuel," University of California Radiation Laboratory—51228, June 19, 1972.

33.  W. J. D. Escher, "The Case for the Hydrogen-Oxygen Car," in The Analog Science Fact Reader (New York: St. Martin's, 1974).

34.  L. W. Jones, "Liquid Hydrogen as a Fuel for Motor Vehicles: A Comparison with Other Systems," Seventh Intersociety Energy Conversion Engineering Conference, 1972, pp. 1364-65.

35.  Roger Billings, Billings Energy Research Corporation, Provo, Utah, personal communication, 1975.

36.  W. D. Van Vorst and J. G. Finegold, "Automotive Hydrogen Engines and Onboard Storage Methods," in Hydrogen Energy Fundamentals: A Symposium Course, ed. T. N. Veziroglu, presented at University of Miami, Miami Beach, Fla., March 3-5, 1975.

37.  L. W. Jones, "Liquid Hydrogen as a Fuel for Motor Vehicles: A Comparison with Other Systems," Seventh Intersociety Energy Conversion Engineering Conference, 1972, pp. 1364-65.

38. W. R. Parrish and R. O. Voth, "Cost and Availability of Hydrogen," in Selected Topics on Hydrogen Fuel, ed. J. Hord, Report NBS IR 75-803, Cryogenics Division, National Bureau of Standards, Boulder, Colo., January 1975, pp. 1-25.

39. W. F. Stewart and F. J. Edeskuty, "Logistics, Economics, and Safety of a Liquid Hydrogen System for Automotive Transportation," presented at the Intersociety Conference on Transportation, Denver, Colo., September 23-27, 1973. American Society of Mechanical Engineers Publication 73-ICT-78.

40. R. E. Billings, "Hydrogen Storage for Automobiles Using Cryogenics and Metal Hydrides," in Hydrogen Energy, ed. T. N. Veziroglu (New York: Plenum, 1975), pp. 791-802.

41. F. E. Lynch, "Metal Hydrides: The Missing Link in Automotive Hydrogen Technology," in Proceedings of the Cornell International Symposium and Workshop on the Hydrogen Economy, ed. S. Linke (Ithaca, N.Y.: Cornell University, April 1975), pp. 408-23.

42. Louis Tellerico, Sandia Corporation, Livermore, Calif., personal communication, 1974.

43. P. Jonville, H. Stohr, R. Funk, and M. Kornmann, "Metal Hydrides: Experimental Methods and Applications to Vehicular Propulsion," in Hydrogen Energy, ed. T. N. Veziroglu (New York: Plenum, 1975), pp. 765-78.

44. C. H. Waide, J. J. Reilly, and R. H. Wiswall, "The Application of Metal Hydrides to Ground Transport," in Hydrogen Energy, ed. T. N. Veziroglu (New York: Plenum, 1975), pp. 779-90.

45. R. D. Witcofski, "Prospects for Hydrogen-Fueled Aircraft," in Proceedings of the Cornell International Symposium and Workshop on the Hydrogen Economy, ed. S. Linke (Ithaca, N.Y.: Cornell University, April 1975), pp. 315-27.

46. "Working Symposium on Liquid-Hydrogen-Fueled Aircraft," National Aeronautics and Space Administration, Langley Research Center, Hampton, Va., May 15-16, 1973.

47. G. D. Brewer, "The Case for Hydrogen-Fueled Transport Aircraft," Astronautics and Aeronautics, May 1974, pp. 40-41.

48. C. Covault, "Fuel Shortages Spur Hydrogen Interest," Aviation Week and Space Technology, December 17, 1973, pp. 38-42.

49. G. D. Brewer, "Advanced Supersonic Technology Concept Study—Hydrogen Fueled Comparison," National Aeronautics and Space Administration, Contract NAS 2-7732, Ames Research Center, Ames, Iowa, January 1974.

50. G. D. Brewer et al., "Study of the Application of Hydrogen Fuel to Long Range Subsonic Transport Aircraft," National Aeronautics and Space Administration, Langley Research Center, Hampton, Va., January 1975.

51. "Development of Liquid-Hydrogen Scramjet Key to Hypersonic Flight," Aviation Week and Space Technology, September 17, 1973, pp. 75-78.

52. "Hypersonic Aircraft by 2000 Pushed," Aviation Week and Space Technology, September 17, 1973, pp. 52-57.

53. J. V. Becker, "New Approaches to Hypersonic Aircraft," paper presented at the Seventh Congress of the International Council of the Aeronautical Sciences, Rome, Italy, September 14-18, 1970.

54. A. L. Nagel and J. V. Becker, "Key Technology for Air-breathing Hypersonic Aircraft," paper presented at the Ninth annual meeting of the American Institute of Aeronautics and Astronautics, Washington, D.C., January 8-10, 1973.

55. C. R. Dyer, M. Z. Sincoff, and P. D. Cribbins, eds., "The Energy Dilemma and Its Impact on Air Transportation," National Aeronautics and Space Administration-American Society for Engineering Education, Langley Research Center, Hampton, Va., 1973.

56. "The Long Range Needs of Aviation," a report of the Aviation Advisory Commission, Washington, D.C., January 1973.

57. H. H. Kolm and R. D. Thorton, "Electromagnetic Flight," Scientific American, October 1973, pp. 17-25.

58. R. A. Hein, "Superconductivity: Large-Scale Applications," Science, July 19, 1974, pp. 211-22.

59. P. E. Glaser, "Satellite Solar Power Station: An Option of Power Generation," Seventh Intersociety Energy Conversion Engineering Conference, 1972.

60. P. Meadows and J. A. DeCarlo, "Hydrogen," in Mineral Facts and Problems, Bureau of Mines Bulletin 650, 1970.

61. Robert Muller, Stanford Research Institute, Menlo Park, Calif., personal communication, 1974.

62. L. R. Brown, "World Food Prospects," Science, December 12, 1975, pp. 1053-59.

63. "Pushing for Priority," Chemical Week, April 3, 1974, p. 15.

64. N. Valery, "Steelmaking with Heat from the Atom," New Scientist, September 13, 1973, pp. 610-15.

65. R. J. Leary and G. M. Larwood, "Effects of Direct Reduction upon Mineral Supply Requirements for Iron and Steel Production," Bureau of Mines Information Circular 8583, 1973.

66. D. J. Blickwede and T. F. Barnhardt, "The Use of Nuclear Energy in Steelmaking," paper presented at the First National Topical Meeting on Nuclear Process Heat Applications, Los Alamos Scientific Laboratory, Los Alamos, N. Mex., October 1-3, 1974.

67. "The New Economics of World Steelmaking," Business Week, August 3, 1974, pp. 34-39.

68. "Coke Oven Control Program Could Cost Bethlehem Steel $40 Million," Air/Water Pollution Report, October 28, 1974, p. 428.

69. "A Search for Clean Coking Processes," Steel Facts, no. 1, American Iron and Steel Institute, Washington, D.C., 1974.

70. A. L. Robinson, "Materials Science: Superconductivity Heats Up," Science, December 28, 1974, p. 1334.

71. A. L. Robinson, "Superconductivity: Surpassing the Hydrogen Barrier," Science, January 25, 1974, pp. 293-96.

72. "Record Heat for Supercold Alloy," Industrial Research, December 1973, p. 20.

73. R. F. Barron, "Liquefaction Cycles for Cryogens," in Advances in Cryogenic Engineering, vol. 17, ed. K. D. Timmerhaus (New York: Plenum, 1972), pp. 20-36.

74. W. D. Metz, "Helium Conservation Program: Casting It to the Winds," Science 183 (January 11, 1974): 59-63.

75. C. A. Price, "The Helium Conservation Program of the Department of the Interior," in Patient Earth, ed. J. Harte and R. H. Socolow (New York: Holt, Rinehart and Winston, 1971).

76. R. A. Hein, "Superconductivity: Large-Scale Applications," Science 185 (July 19, 1974): 211-22.

77. K. Imatani and K. D. Timmerhaus, "Current Status of Cryogenic and Air-Blast Food Freezing Systems," in Advances in Cryogenic Engineering, vol. 17, ed. K. D. Timmerhaus (New York: Plenum, 1972), pp. 137-46.

78. E. G. Valdez et al., "Use of Cryogens to Reclaim Nonferrous Scrap Metals," Bureau of Mines, Report of Investigations no. 7716, 1973.

79. B. Landau, "From Sledgehammers to Shredders: Scrap Processing Technology Advances," Phoenix Quarterly, Institute of Scrap Iron and Steel, Inc., Fall 1975, p. 1.

# 8

## HYDROGEN SAFETY

### COMPLEXITY OF THE SAFETY QUESTION

The safety of hydrogen compared with commonly used fuels is a complicated question with no clear-cut answers. Among those in the technical community who have no direct contact or operating experience, there is widespread belief that hydrogen is a dangerous substance that burns and explodes readily. Hydrogen economy enthusiasts often argue, however, that if it is used properly, hydrogen is really no more dangerous than the familiar fuels—methane (natural gas), gasoline, diesel, propane, and kerosene (jet fuel) [1, 2].

Neither attitude is completely correct, because the hazards of any substance depend on the nature and environment of use as well as on whether the moment is routine or extraordinary. It can be said, however, that since the physical and chemical properties of hydrogen are quite different from commonly encountered substances, the hazards associated with hydrogen appear to be far more influenced by circumstances than for other fuels.

This chapter examines the physical and chemical aspects of hydrogen safety, and also social aspects, such as the importance of peoples' perceptions about hydrogen safety and the distinction between voluntary and involuntary exposure to risk. It will be seen that blanket statements about the safety of hydrogen relative to other substances are generally misleading.

### PROPERTIES OF HYDROGEN

The physical and chemical properties of hydrogen, methane, gasoline, and jet fuel relevant to safety are shown in Table 8.1. The most important points about hydrogen to note are the following:

TABLE 8.1

Properties of Hydrogen and Other Common Fuels Relevant to Safety

| Property | Hydrogen ($H_2$) | Methane ($CH_4$) | Methanol ($CH_3OH$) | Gasoline | Jet Fuel (JP-4) |
|---|---|---|---|---|---|
| Temperature of liquid at normal boiling point | 20.3°K | 112°K | 338°K | a | a |
| Heat of vaporization | 0.45 MJ/kg | 0.51 MJ/kg | 1.1 MJ/kg | — | — |
| Density of hydrogen gas (at normal boiling point) relative to the density of air (at STP) | 1.03 | 1.38 | — | — | — |
| Density of gas relative to the density of air (at STP) | 0.070 | 0.55 | — | — | — |
| Diffusion coefficient | 0.63 cm²/sec | 0.2 cm²/sec | — | 0.08 cm²/sec | — |
| Flammability limits in air (by volume) | (4.1–74)% | (5.3–15)% | (6.0–37)% | (1.5–7.6)% | (0.8–5.6)% |
| Detonability limits in air (by volume) | (18–59)% | (6.3–14)% | — | — | — |
| Ignition temperature | 850 K | 807 K | 700 K | 530 K | 522 K |
| Ignition energy | 20 μJ | 300 μJ | — | 250 μJ[b] | — |
| Flame temperature | 2400 K | 2190 K | — | — | — |
| Flame velocity | 2.75 m/sec | 0.37 m/sec | 0.41 m/sec | < 0.3 m/sec[b] | — |
| Quenching distance | 0.06 cm | 0.23 cm | — | > 0.25 cm[b] | — |
| Emissivity of flame | 0.10 | 1.0 | — | — | — |
| Heat of combustion | 120 MJ/kg | 50 MJ/kg | 20 MJ/kg | 44 MJ/kg | 43 MJ/kg |
|  | 8.5 GJ/m³ | 21 GJ/m³ | 16 GJ/m³ | 31 GJ/m³ | 34 GJ/m³ |

[a]Not applicable; these fuels are not single molecular species, and their components exhibit a wide range of boiling points.
[b]Estimated in reference 2.

1.  Liquid hydrogen is a very cold substance (-452°F).

2.  Gaseous hydrogen at room temperatures has a very low density compared with that of air and thus is very buoyant [1, 3, 4].

3.  Gaseous hydrogen diffuses through air very quickly [1, 3].*

4.  Hydrogen is flammable in air over a wide range of concentrations, much more so than methane (natural gas) [1, 3, 4].

5.  When confined, hydrogen is detonatable over a wide range of concentrations [1, 3], but when unconfined it is difficult to detonate.

6.  Hydrogen can be ignited by a very small amount of energy.

7.  Hydrogen flames travel much faster than methane flames [1, 3].

8.  Hydrogen flames are nearly invisible [1, 3].


## HYDROGEN BEHAVIOR

Liquid hydrogen spilling from a large rupture in a large storage tank would be dramatic: The liquid hydrogen, its temperature well below the usual ambient temperature, would quickly draw heat from the surroundings and vaporize rapidly. The cold hydrogen gas would occupy about 50 times the volume it did as a liquid. As the cold gas continued to draw heat from the surroundings and to warm up, water vapor in the chilled surrounding air would condense or freeze and form a dense white cloud of fog. Some oxygen and nitrogen in the air would also condense or freeze. Nearby objects would contract as they underwent rapid cooling, giving off noises. As the volume of hydrogen gas warmed it would continue to expand and, when it became less dense than the chilled air surrounding it, it would rise buoyantly above the spill. Simultaneously, individual molecules of hydrogen would diffuse from the volume of gas in all directions. If the spill were in an open environment, such as out of doors or under a simple canopy, the hydrogen would disperse quickly. But it if occurred in a building, the hydrogen could be trapped and the large change in hydrogen volume upon vaporization would force air from the environment. The extent and rate of dispersal and the nature of the nearby objects would determine whether the hydrogen would ignite. Figure 8.1 depicts the observed dispersal of flames from a small spill of liquid hydrogen that has ignited [5]; note that in two seconds the flames are nearly all dis-

---

*Buoyancy and diffusivity are different phenomena: buoyancy describes the rate that a packet or cloud or gas would move intact; diffusivity describes the rate that an individual molecule would penetrate through a volume of another gas. Both parameters are required to describe how rapidly a cloud of hydrogen would disperse in air.

FIGURE 8.1

Maximum Vertical Cross-Sections of Flames Produced at Various
Time Intervals Following Spillage of 89 Liters of Liquid Hydrogen
on a Gravel Surface
(horizontal scale = vertical scale)

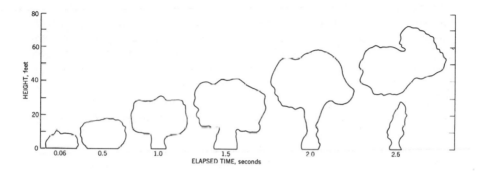

Source: Reference 5.

persed upward. Extreme care must be exercised, however, in drawing inferences from this figure about the behavior of a large spill. *
If liquid hydrogen were to leak from a small orifice, the effects would
be less dramatic than in a large rupture spill, but they would still be
impressive, especially since a stream leaking from a small orifice
often ignites because of static electric discharges [6].

The description above illustrates that blanket statements about
hydrogen safety are inadequate because the final result of the spill
depends upon the nature of the rupture, the degree of confinement,
the nature of the actual nearby objects, and whether combustion is
initiated. With the present state of knowledge about hydrogen and its
accidents, discussions of its safety usually assume the form of "on
the one hand this, but on the other hand that." Some of these kinds of
discussions follow.

The wide range of flammability of hydrogen in air suggests that
a leak of hydrogen is more likely to result in a fire than is a leak of
methane. For example, hydrogen can ignite when the atmosphere is

*Recall that a railroad tank car holds about 28,000 gallons of liq-
uid hydrogen. This is more than 1,000 times larger than the spill
shown in Figure 8.1.

50 percent hydrogen while, at the same percentage, methane cannot. After a leak, however, the fuel concentration builds up from zero, and ignition (if an ignition source is present) is most likely to occur when the concentration first reaches the lower flammability limit [1, 3, 4]. If the concentration passes this level without igniting, then most likely there are no ignition sources present and ignition will not occur directly. Consequently, the lower flammability limit is really most important irrespective of the total range [1, 3, 4]. Table 8.1 shows that in this respect hydrogen is similar to methane, a fuel that most people regard as acceptably safe.* The ignition energy parameter is also relevant here, however. Table 8.1 shows that hydrogen can be ignited by only about one-fifteenth the energy required to ignite methane, which greatly increases the likelihood of hydrogen ignition. For example, synthetic fibers in clothes cause electrostatic sparks, many of which are imperceptible to the wearer but which are energetic enough to ignite hydrogen [7].

The wide detonatability limits of hydrogen lead to an analysis similar to the above. Yet in often mentioned experiments in 1960, the consulting firm of Arthur D. Little, Inc., was unable to detonate hydrogen after creating spills of liquid hydrogen up to 5,000 gallons in volume [8]. Other workers, however, insist that hydrogen is readily detonated and report accidents [9]. The discrepancy in experience appears to be traceable to the presence of walls and other obstacles that can reflect pressure waves.

When pure hydrogen (with no flame colorant added) burns, the flame is nearly invisible in daylight because little energy is radiated [1]. While the invisibility of the flame makes the fire difficult to locate and fight, firefighters can get very close to the flame (assuming they know where it is) without injury. Since it is also difficult to feel warmth from the flame, a person can easily move right into the flame and be burned; but surrounding objects do not heat up and ignite unless touched by the flame directly. In normal fires, one of the major causes of fire spreading and injury is the large amount of energy radiated by oxidizing carbon atoms. Thus, the lack of flame luminosity in hydrogen fires can be both a help and a hindrance.

The energy radiated from fires above pools of liquid hydrogen and liquid methane of the same area and volume would be nearly identical because the lower emissivity of hydrogen is offset by the more

---

*This attitude is held in spite of numerous accidents each year. For example, there were more than 39 deaths and 218 injuries in the first half of 1973 alone from accidents with natural gas, many of them in residences [10]. This gives a hint of the kind of accident rate people accept for voluntary exposure.

rapid rate of combustion [7]. Of course, this also means that the available fuel is more quickly exhausted in the hydrogen fire. However, owing to its rapid rate of evaporation, pools of liquid hydrogen are not likely to form except after a very large spill. A liquid-hydrogen-fueled airplane that crashed on takeoff would probably result in such spills.

Hydrogen's wide range of flammability and low ignition energy are offset by its great tendency to disperse from the scene of a leak or a spill because of its buoyancy and high rate of diffusion (see Figure 8.1). An accident in which a tank truck full of liquid hydrogen rolled over with no resulting fire and no harm to the driver is sometimes cited in the literature [11]. The hydrogen evaporated and dispersed quickly upward from the crashed truck. Because there have been few accidents to date, the statistics on hydrogen tank truck accidents are too few to stress this evidence. No doubt the statistics of gasoline tank truck accidents also contain reports of comparable harmless events; they also contain reports of truly devastating events. In contrast with hydrogen, when gasoline is spilled, vapors heavier than air spread in a wide layer near the ground. This greatly increases the hazardous area. Thus, the tendency of hydrogen to disperse much more rapidly than other fuels, even gaseous methane, is a large point in its favor.

Metal hydrides offer certain safety advantages over liquid or gaseous storage and storage and distribution of hydrogen because heat must be supplied to expel hydrogen from the metal hydride. Consequently, metal hydrides cannot leak to any appreciable extent (only a tiny amount randomly diffusing out of the hydride). If the hydrides were used as a fuel tank in an automobile, for example, an accident could not directly cause release of hydrogen. However, should a fire start for other reasons and heat the hydride sufficiently, then hydrogen would be released to add to the fire. Depending upon the actual metal hydride involved and the specifics of the situation, the release could be self-sustaining or could be self-extinguishing because of a deficiency of heat. However, since powdered metals often burn (especially magnesium, which is pyrophoric), the metal carrier may itself pose a hazard.

## EXPERIENCE IN THE SPACE PROGRAM AND THE HINDENBURG

Frequent reference is made to the excellent safety record maintained by NASA in its use of vast quantities of liquid hydrogen in the U.S. space program [1, 3, 4, 11-14]. Unfortunately, this excellent record should not be extrapolated to the large-scale use of liquid hydrogen by the general populace because of several important differences:

1.  NASA personnel handling hydrogen are specially chosen and trained [14].
2.  Hydrogen use in the space program has been restricted to rigorously controlled out-of-doors environments.
3.  Extreme concern about the reliability of systems has dominated the entire space program, thereby minimizing the chance of accidental release of hydrogen.
4.  Hydrogen safety procedures to protect both personnel and expensive, often unique, hardware have received major emphasis.

Undoubtedly the NASA experience offers encouragement about the safe use of liquid hydrogen and should serve as a point of departure for the development of more generally applicable future hydrogen safety procedures [11].  Nevertheless, the NASA experience is simply inadequate to serve as a basis for decision making concerning more general aspects of a hydrogen economy in which the substance would be generally in unskilled hands.

An informal club called the $H_2$indenburg Society has arisen among the fraternity of people who advocate the hydrogen economy. These people believe that the general public associates hydrogen with the disaster in which the dirigible Hindenburg was destroyed by fire at Lakehurst, New Jersey, in 1937; they call this association the "Hindenburg Syndrome."  Although they were designed to use helium for buoyancy, the large German airships of the 1930s were filled with gaseous hydrogen because the United States would not sell helium to Germany [1, 15].  The Hindenburg, violating safety rules, approached Lakehurst through a thunderstorm and caught fire and burned while attempting to dock [1, 15].  A large technologically related disaster for its time, the Hindenburg accident killed 36 passengers and 22 crewmen [15], but it is usually forgotten that 65 people on board survived [1].*  The number of people killed in the Hindenburg is small compared with the number killed in crashes of modern aircraft.

The importance of the Hindenburg accident to the hydrogen economy lies less in what it reveals about the actual safety of hydrogen than in what writers, journalists, and movie makers attempt to make of it.  The authors question whether the general public really does associate the hazard of hydrogen with the Hindenburg.  Instead, we sus-

---

*Publicists little note the successful career of the hydrogen-filled Graf Zeppelin, which made regular and safe crossings of the Atlantic from 1928 until her retirement in 1937.

pect that this association is common mainly among people with technical training because they are most apt to recall their experiments in high school or college chemistry courses. However, the movie Hindenburg may seriously distort public attitudes.

## SPECIAL HAZARDS

Two special hazards associated with hydrogen are not as widely known as its flammability. One is the extreme cold of liquid hydrogen, which can severely damage or kill living tissue. The other is hydrogen embrittlement of materials, which can cause storage, distribution, and utilization facilities or devices to fail.

Since contact with an extremely cold substance produces a sensation much like that of a heat burn, wounds inflicted by cold substances are also called "burns." The likelihood of liquid hydrogen spilling and splashing on many people to cause such burns is relatively small because of hydrogen's rapid rate of evaporation. Only in a massive spill, in which the environment was so depleted of heat energy that evaporation was impeded, would there be the likelihood that people would come into direct contact with the liquid. However, damage to living tissue could also result from contact with the greatly chilled objects or air surrounding a spill. Experience with cryogenic liquids is still rare, and the hazards are not yet appreciated by the general public.

Hydrogen environment embrittlement (see Chapter 6) can cause hydrogen storage, distribution, or utilization devices to fail [17-19], thereby releasing hydrogen, which might ignite and burn. In addition, the failure of a mechanical part could cause an accident that might have consequences even more serious than a hydrogen fire or explosion. Because hydrogen environment embrittlement is most pronounced in high-strength steels exposed to high-pressure, high-purity hydrogen near room temperatures [16-18], many hydrogen-related activities, such as liquid handling, would be immune from embrittlement. Nevertheless, it is now generally recognized that a crucial part of guaranteeing occupational and public safety in the face of hydrogen usage will be careful evaluation of the materials at relevant values of temperature, pressure, and hydrogen purity. It is generally believed that engineering solutions can be found to assure a reasonable degree of safety [17]. It might be found, however, that certain geophysical hazards make hydrogen environment embrittlement more dangerous in some areas than in others. In particular, large hydrogen gas pipelines traversing seismically active areas—such as California or Alaska—might prove especially dangerous.

## SUMMARY OF PHYSICAL AND CHEMICAL
## ASPECTS OF SAFETY

There are too many compensating aspects of the physics and chemistry of hydrogen to make blanket statements about its actual safety. Instead, the safety question must be addressed and answered on a situation-by-situation basis. In each case, the important parameters that affect the outcome of the evaluation include:

1. The geometry and degree of confinement of the environment in which the hazards exist
2. The proximity, number, strength, and duration of ignition sources
3. The temperature, pressure, and purity of the hydrogen and the nature of the materials exposed to it
4. The intelligence, experience, training, skill, and regimentation of people exposed to the hazards
5. Consideration of failure events, as well as routine circumstances

## PERCEPTIONS OF SAFETY HAZARDS—A KEY
## TO POLICY

Public perceptions of the safety of hydrogen may be one of the major obstacles to a transition to a hydrogen economy. A key to policy making lies in recognition of the distinction between two different types of perceptions and their implications: familiarity with "real" hazards, as opposed to "nonreal" or imagined hazards. Decision makers in both the public and private sectors must take into account both types of perceptions.

Perceptions of "real" hazards usually have their basis in familiarity with technical facts derived from physics, chemistry, and the conditions of the environment of exposure. Certainly, many "real" hazards exist in the utilization of hydrogen, but these hazards are not absolute; instead, they are relative to the specific conditions of use.

Perceptions of "nonreal" hazards are based on incomplete knowledge and can be further influenced by many factors. Although there are many different theories on attitudes and attitude formation, each arising from different psychological theories of personality, it is generally recognized in most theories that attitudes are based in part on "beliefs." A person's beliefs are affected by a seemingly endless list of factors including experience, education, age, income, sex, geographic location, ethnic group, religion, and exposure to news media. Since the actual safety of hydrogen is so variable and so dependent on the specific conditions of its use, the factor of education seems to be especially important. A fairly high level of knowledge about the technology would be required to understand all the relevant qualifying conditions.

Individuals with the power and responsibility to make decisions generally have access to technical information superior to that traditionally available to the general public. Since comprehensible technical information has generally been lacking, the public has been forced to rely on the downward trickle of bits and pieces of information, many of which are easily misunderstood. This information frequently contains inconsistencies or contradictions, particularly in "gray" areas where there is disagreement among experts.

Three factors seem especially relevant in the conditioning of the public's perceptions about the safety of hydrogen: the newness of many aspects of the technology; the usual lack of understandable technical information for the general public; and the apparently contradictory experience and statements about hydrogen safety. The combination of these three factors could easily create an atmosphere within which the public would be opposed to hydrogen, especially since there will be a tendency to publicize mainly the hazardous aspects. As a result, the public is not likely to base its perceptions of hydrogen safety on a "real" basis. This problem is confounded by the long period of transition to the hydrogen economy during which there is a turnover in the population composing the public. Decision makers will have to respond to public attitudes no matter what type of information lies behind them. In particular, it would matter little whether negative attitude toward hydrogen were based on "real" or "nonreal" perceptions of hazards; both could have the same impact on the ultimate acceptance or rejection of the hydrogen economy.

The increase in both the number and size of citizen action groups over the past decade indicates heightened public interest in environmental and consumer affairs. These groups have increasingly been able to influence legislation and corporate decisions. Unless public apathy sets in before the hydrogen economy begins to emerge, it will be crucial to obtain general acceptance prior to implementation of major hydrogen facilities. If acceptance fails to develop, attempts to block or modify implementation should be expected.

## VOLUNTARY VERSUS INVOLUNTARY EXPOSURE TO HAZARDS

Exposure to hydrogen will be involuntary once a transition to a hydrogen economy begins. Consequently, pressures for stringent regulation regarding hydrogen use can be expected to develop.

Voluntary exposure to hazards results from individual choice. The most clear-cut examples of voluntary exposure are found in recreational activities such as skiing, auto racing, scuba diving, and playing football. Before participating in these activities the individual has generally decided that the risk is acceptable compared to the rewards. However, involuntary exposure to hazards occurs when the decision is beyond an individual's reasonable control. Natural conditions that create involuntary exposure include hurricanes, earthquakes, floods, and tornados. The technology of modern civilization also creates involuntary exposure to such hazards as air pollution, water pollution, gasoline tank trucks on the highways, and airplanes flying overhead. Technologically induced involuntary exposure to hazards is the result of governmental or corporate decisions beyond individual control.

Recent years have seen a trend toward increasingly broad and stringent regulation of activities that create involuntary exposure to hazards. The increased governmental regulation in safety devices and food and drug testing are examples of this trend. The combination of the likely misconceptions about hydrogen safety and involuntary exposure are likely to result in tough new restrictions upon hydrogen technologies. This control may even be more restrictive than that which has been applied to other fuels or energy technologies, merely because the hydrogen economy is starting free of established major pro-hydrogen interests, but with a large number of interests that stand to lose if hydrogen use becomes established. Thus, ironically, because stringent regulations can lead to safe technology and practice, the hydrogen economy could conceivably prove to be safer than the existing state of affairs.

REFERENCES

1.  D. P. Gregory et al, "A Hydrogen-Energy System," American Gas Association, Alexandria, Va., August 1972.

2.  W. F. Stewart and F. J. Edeskuty, "Logistics, Economics, and Safety of a Liquid Hydrogen System for Automotive Transportation," presented at the Intersociety Conference on Transportation, Denver, Colo., September 23-27, 1973. American Society of Mechanical Engineers Publication 73-ICT-78.

3.  L. T. Blank et al., "A Hydrogen Energy Carrier," Systems Design Institute, National Aeronautics and Space Administration-American Society for Engineering Education, Johnson Space Center, Houston, Texas, 1973.

4.  "Hydrogen and Other Synthetic Fuels," a summary of the work of the Synthetic Fuels Panel, prepared for the Federal Council on Science and Technology R&D Goals Study, September 1972.

5. M. G. Zabetakis and D. S. Burgess, "Research on the Hazards Associated with the Production and Handling of Liquid Hydrogen," Bureau of Mines, 1961.

6. Thomas Goodale, Poulter Laboratories, Stanford Research Institute, Menlo Park, Calif., personal communication, 1974.

7. D. B. Chelton, "Safety in the Use of Liquid Hydrogen," in Technology and Uses of Liquid Hydrogen, ed. R. B. Scott, W. H. Denton, and C. M. Nicholls (New York: Macmillan, 1964), pp. 359-78.

8. L. H. Cassutt, F. E. Maddocks, and W. A. Sawyer, "A Study of the Hazards in the Storage and Handling of Liquid Hydrogen," in Advances in Cryogenic Engineering, vol. 5 (New York: Plenum, 1959), pp. 55-61.

9. Erwin Capener, formerly of Stanford Research Institute, Menlo Park, Calif., personal communication, 1974.

10. "When Gas Pipelines Blow Up," Business Week, August 4, 1973, p. 50.

11. P. M. Ordin, "A Review of Hydrogen Accidents and Incidents in NASA Operations," Ninth Intersociety Energy Conversion Engineering Conference, 1974, pp. 442-53.

12. F. A. Martin, "The Safe Distribution and Handling of Hydrogen for Commercial Application," Seventh Intersociety Energy Conversion Engineering Conference, 1972, pp. 1335-41.

13. J. R. Bartlit, F. J. Edeskuty, and K. D. Williamson, Jr., "Experience in Handling, Transport, and Storage of Liquid Hydrogen —The Recyclable Fuel," Seventh Intersociety Energy Conversion Engineering Conference, 1972, pp. 1312-15.

14. Hydrogen Safety Manual, Advisory Panel on Experimental Fluids and Gases, National Aeronautics and Space Administration, Lewis Research Center, Cleveland, Ohio, NASA Technical Memorandum TM X-52454, 1968.

15. J. Toland, The Great Dirigibles—Their Triumphs and Disasters, rev. ed. (New York: Dover, 1972).

16. A. W. Thompson, "Structural Materials Use in a Hydrogen Economy," paper 37c presented at the annual meeting of the American Institute of Chemical Engineers, Washington, D.C., December 3, 1974.

17. R. P. Jewett et al., "Hydrogen Environment Embrittlement of Metals," National Aeronautics and Space Administration Report CR-2163, March 1973.

18. H. H. Johnson and A. J. Kumnick, "Hydrogen and the Structural Integrity of Alloys," in Hydrogen Energy, ed. T. N. Veziroglu (New York: Plenum, 1975), pp. 1043-56.

In technology assessment it is important to consider not only the particular technology in question but also technologies that may significantly compete with it or complement it. Only when the technology is seen in context is there a reasonable chance to gauge the impacts meaningfully. Sometimes it is found that the technology studied is susceptible to preemption by cheaper, more convenient, or more socially acceptable technologies; alternatively, the advent of complementary technologies often makes deployment more likely.

Because there are so many end-uses of energy and so many options for developing new technologies of energy supply, transformation, storage, distribution, and end-use, we must place the hydrogen economy concept in perspective. The following two chapters discuss competing and complementing technologies.

# 9

## ENERGY CARRIER, DISTRIBUTION, AND STORAGE ALTERNATIVES TO HYDROGEN

ENERGY CARRIER

Hydrogen is not used as an energy carrier today in any significant way. To gain a foothold, hydrogen therefore must compete with all the common chemical fuel substances, but especially with the many products derived from petroleum. In the short run, petroleum fuels have a substantial advantage over hydrogen because they are readily obtained from natural deposits and can be converted into refined products tailored to existing end-uses with relatively high energy efficiency. Moreover, since refined petroleum products are naturally in a liquid form at room temperatures, they are more easily transported and stored than is hydrogen, which is in a liquid form only at very low temperatures ($-423\,°F$). These advantages are reflected in hydrogen having a higher price on a unit energy basis than petroleum products; consequently, hydrogen faces stiff competition from traditional petroleum resources for probably two decades or more.

Even as domestic reserves of petroleum diminish, hydrogen will face competition from synthetic hydrocarbon fuels derived from coal, oil shale [1], and organic wastes. Petroleum companies and the fuel-supply sector in general have a strong incentive to develop and produce synthetic fuels essentially identical to current petroleum products because this enables them to protect their very large investments in existing refining, storage, and distribution facilities [1]. Moreover, synthetic fuels that are chemically nearly the same as the traditional hydrocarbon fuels spare customers the discomfort of change [1]. The petroleum companies themselves can progressively absorb the necessary cost of change and can confine most of it to that portion of the business devoted to resource extraction. This would be accomplished by producing a synthetic crude oil that could simply be blended with remaining supplies of natural crude at the refinery [1].

This ability to "roll in" synthetic crude gradually is responsible for the attention major petroleum companies are giving to coal and oil shale; it is also responsible for their notable lack of interest in hydrogen. Within even the most distant planning horizons of concern to these companies (year 2000 or so), "syncrudes" from coal and oil shale are much more appealing than hydrogen because they are more compatible with company expectations regarding the future total world petroleum picture [1]. Consequently, there is considerable momentum behind development of synthetic fuels essentially molecularly identical to conventional fuels.* Hydrogen must compete with such fuels within this context at least until the early twenty-first century.

Distributors of natural gas and the gas utilities demonstrate an attitude similar to that of the petroleum companies about the development of synthetic gaseous fuels. They have placed considerable emphasis on development of technologies to produce methane as a substitute natural gas (SNG) from coal or lignite (a low-grade form of coal). Since natural gas as delivered to consumers is nearly pure methane, gas companies could progressively phase in SNG without compromising their considerable investment in existing facilities; again, the consumer would not be caused any discomfort or even be made aware of the substitution, except by the cost of the energy. Thus, hydrogen must compete not only with natural resources or methane but also with SNG for a role in the gas market.

In the very long run, of course, when petroleum, natural gas, and coal resources are not available, hydrogen would have less competition. Nevertheless, during the process of a transition to hydrogen, its ability to compete with these other energy sources and forms is crucial.

Hydrogen must also compete with electric power. At the point of use, not only is electricity a very high-quality and efficient form of energy; but it is environmentally clean, and electrical end-use devices are quiet. Moreover, institutionally, electricity has many of the same attributes described for synethetic crude oil and SNG. In particular, practically any energy source can be utilized to generate electricity—including potential sources not yet in general use. Many electric generating technologies are potentially available, including the following (those not yet in use to a major extent are marked with an asterisk):

Hydroelectric
Fossil-fuel-fired boilers

_____

*Although there was marked hesistancy in 1974-75 because of a lack of firm national energy policy to define the "rules of the game" [1].

Nuclear [2]
   Conventional light-water fission reactors
   Gas-cooled high-temperature fission reactors
   Fission breeder reactors*
   Fusion reactors*
Fuel cells*
Geothermal [3]*
Organic-waste-fired boilers [3]*
Solar [3]*
   Thermal
   Photovoltaic
   Ocean thermal
   Windpower
   Biomass plantations

Again, the effect of change in the adoption of new resources or technologies could be confined to the electric generation portion of the electric system, leaving the distribution network and the consumer unaffected. In this context, the electric power network is already in place and possesses much resiliency with respect to ultimate generation methods. This ensures that hydrogen will have to compete with the already familiar, ubiquitous, clean, flexible, electric power system. The largest single constraint on increased use of electricity is the technical difficulty and the high investment costs of expanding the system [4]. At present, the most critical nonfinancial constraints are control of air pollution emissions at fossil-fuel plants, nuclear power plant siting and waste disposal, and obtaining overhead transmission line corridors. Use of electricity is difficult in nonrail transportation applications, however.

Technologies that complement hydrogen used as an energy carrier are essentially any that are themselves made more effective by the use of hydrogen or those that make the use of hydrogen cheaper or more convenient. In the former category are the many forms of solar energy, which need energy storage mechanisms to balance variation in output arising from predictable cycles (such as day and night) or from fluctuating weather. Indeed, the exploitation of ocean thermal gradients [5, 6] to generate energy may only be feasible if the energy can be transported to shore in the form of hydrogen [5]. As a simple matter of efficiency, however, once energy was in the form of hydrogen, there would be an incentive to sell it directly rather than to pay the energy loss penalty of reconversion to electricity.

The category of technologies that could enhance the likelihood for hydrogen use as an energy carrier includes:

1. Simpler and cheaper cryogenic liquid production, storage, and distribution

2. Steels immune to hydrogen environment embrittlement

3. Lightweight metal hydrides composed only of abundant materials

4. Low-cost, high-efficiency catalysts for electrolyzers and fuel cells using only abundant materials

5. Corrosion-resistant metal alloys suitable for high-temperature thermochemical water-splitting containment

## ENERGY DISTRIBUTION

The large-scale distribution of hydrogen is most sensible either in gas transmission pipelines [9] or, possibly, by sea in large liquid-hydrogen tank ships. As a substitute for the delivery of electricity from remote generation sites, underground hydrogen pipelines would compete not only with ordinary overhead and advanced underground high-voltage electric transmission but also with future cryoresistive and superconducting electric transmission technologies [10, 11] (generally placed underground).* Cryoresistive transmission lines that use liquid nitrogen coolant are under development by several major corporations, including General Electric, one of the large traditional suppliers to electric utilities. Superconducting transmission lines are under development by the Linde Division of Union Carbide Corporation and under government sponsorship at Los Alamos Scientific Laboratory, Brookhaven National Laboratory, Stanford University, and elsewhere.

Superconducting electric transmission is most relevant to the hydrogen economy concept because it may both compete with hydrogen as an energy carrier and spur the use of some hydrogen. Most plans for superconducting transmission lines call for a niobium (Nb) or niobium-tin alloy superconductor cooled by liquid helium [10]. An alloy of vanadium and gallium ($V_3Ga$) has also been discussed. However, the recent discovery that niobium-germanium ($Ng_3Ge$) possesses a superconducting transition temperature (about 23° K) above the normal boiling point (20.4°K) of liquid hydrogen [12, 13] has lent great encouragement to the hope of developing a transmission line that could use liquid hydrogen as a coolant.

However, there may be significant resource limitations to the scale of deployment of superconducting transmission lines. First, unless the federal helium conservation program is reinstated, or the

---

*At low temperatures, nonsuperconductors exhibit reduced resistance to the flow of electricity and hence dissipate less energy during transmission. Superconductors exhibit no resistance at all.

use of helium in electric arc welding and other dissipative uses diminishes, a shortage of helium might constrain deployment of helium-cooled superconducting technologies [13]. The expected requirement for niobium should not be a limitation, especially because the new $Nb_3Ge$ high-temperature superconductor can be made in thin films by evaporation techniques, thereby greatly reducing the niobium requirement. However, there may be an absolute shortage of germanium resources to support both the technology of superconducting transmission lines and the semiconductor industry, which is presently the largest user of germanium. Table 9.1 shows the resource demands and limitations.

Several proposals have been advanced for the joint delivery of liquid hydrogen and electricity through an "energy pipeline" based upon superconducting technology [14, 15]. Although realization of this concept lies in the distant future and depends on further advances in superconductor technology, this approach might ultimately be used, for example, to deliver energy derived from Wyoming coal to Chicago [15]. If this very speculative idea proves feasible, a large quantity of liquid hydrogen could be delivered.

Distribution of vast quantities of hydrogen in liquid form makes economic sense only when it is not feasible to transport gaseous hydrogen by pipeline for liquefaction close to the final user. * The Middle Eastern oil fields are potentially a large source of liquid hydrogen delivered by tank ship to the United States. Currently, most of the natural gas production concomitant with Middle Eastern oil production is burned or "flared-off" as waste because local demand is inadequate to justify collection and distribution. Instead of wasting the gas, however, three methods have been suggested for turning a profit on the gas export in sales: liquefaction and export as LNG, conversion to methanol for export, and conversion to hydrogen (by reforming) and liquefaction for export. Today some LNG is imported from Algeria. When the distance of transportation exceeds 4,000 miles, the net energy efficiency of the total system is reported to be higher with methanol than with LNG [16]. Not long ago, however, an analysis performed by a group of Saudi Arabians suggested that the conversion to liquid hydrogen is as economical as conversion to methanol [17]. There is some doubt, however (even among proponents of the hydrogen economy), that this analysis is correct.

---

*The present long-distance delivery of liquid hydrogen by train reflects the lack of a hydrogen pipeline network rather than the long-term economic competitiveness of the method.

## TABLE 9.1

Superconducting Electric Transmission Line Projected Materials Needs and Reserve Estimates

| Element | Factor per MVA$^a$-mile | Year 2000 (1.4 × 10$^6$ MVA−miles) | Year 2030 (6.5 × 10$^6$ MVA−miles) | U.S. Reserves[b] of the Element | World Reserves[b] of the Element |
|---|---|---|---|---|---|
| Helium | 320 SCF | 4.5 × 10$^8$ SCF | 2.0 × 10$^9$ SCF | 35 × 10$^9$ SCF (stored) | — |
| Niobium | 1.1 lb | 1.6 × 10$^6$ lb | 7.2 × 10$^6$ lb | 10$^8$ lb | 10$^{10}$ lb |
| Germanium | 0.2 lb | 0.3 × 10$^6$ lb | 1.3 × 10$^6$ lb | 0.8 × 10$^6$ lb | 3 × 10$^6$ lb |
| Tin | 0.32 lb | 0.45 × 10$^6$ lb | 2.1 × 10$^6$ lb | <10.0 × 10$^6$ lb | 10$^{10}$ lb |
| Gallium | ~0.2 lb | ~0.3 × 10$^6$ lb | ~1.3 × 10$^6$ lb | 6.0 × 10$^4$ lb | — |

[a]Million volt amperes. This unit of power, common in electric transmission circles, is equivalent to a megawatt.

[b]"Reserves" must be distinguished from "resources." Resources are a measure of the known or speculated quantity of an element present in the earth's crust, while reserves are the portion of these resources that can be recovered using known technology at present economic conditions. Resource estimates change as new geological information is obtained, and reserve estimates change when technology or market prices change as well.

## ENERGY STORAGE

As a chemical means to store energy, hydrogen would have to compete with traditional and synthetic forms of fossil fuels, which all have superior storage and transfer characteristics; this includes cryogenic liquid natural gas (LNG) (nearly pure methane). The normal boiling point of LNG is 112°K—considerably closer to ambient temperature (293°K) than liquid hydrogen at 20.4°K, and the volumetric energy content of LNG is greater than that of liquid hydrogen. Moreover, LNG storage is far simpler than for liquid hydrogen. Tanks to store LNG can be made easily in flat-bottomed cylindrical shapes using either concrete lined with plastic foam insulation or double-walled, insulated metal [7]. These are easier and cheaper to build than the double-walled, heavily insulated spherical vessels required for liquid hydrogen.

For short-term storage of energy, as might be employed by either an electric utility [8] or in some solar energy collection systems, there are many technologies competing with hydrogen technologies. Some of these competitors have not been proved feasible on the scales required, however. A list of these energy storage alternatives follows (those that are still speculative are marked with an asterisk):

Pumped hydroelectric
Compressed air*
Magnetic fields in superconducting solenoids*
Advanced concept flywheels*
Underground storage of hot water*
Advanced high-energy and power density batteries*

Because most of these energy storage technologies are as speculative as hydrogen system storage, there is a good chance that hydrogen system storage will prove at least as effective and thus sometimes will be adopted. Three key factors affecting the future choice of options for energy storage are geographic suitability, comparative economic cost, and comparative net round-trip energy system efficiency.* The latter will also be reflected somewhat in the relative cost, of course, but as energy conservation assumes greater importance, the net energy efficiency will also assume increasing importance.

The four most important technologies that complement hydrogen energy storage systems are cryogenics, metal hydrides, electrolysis, and fuel cells. Significant strides, especially in the area of total costs,

---

*For example, electricity → conversion → storage → conversion → electricity.

in any or all of these could tip the balance toward hydrogen systems and away from alternative systems.

## REFERENCES

1. E. Dickson et al., "Impacts of Synthetic Liquid Fuel Development for the Automotive Market," Stanford Research Institute, Menlo Park, Calif., for the Energy Research and Development Administration, Washington, D.C. (in preparation).

2. D. J. Rose, "Nuclear Eclectic [sic] Power," Science 184, no. 4134 (April 19, 1974): 351-59.

3. E. E. Hughes et al., "Control of Environmental Impacts of Advanced Energy Sources," Environmental Protection Agency Report EPA 600/2-74-002, March 1974.

4. "Utilities: Weak Point in the Energy Future," Business Week, January 20, 1975, pp. 46-54.

5. A. Lavi and C. Zener, "Plumbing the Ocean Depths: A New Source of Power," IEEE Spectrum, October 1973, pp. 22-29.

6. W. D. Metz, "Ocean Temperature Gradients: Solar Power from the Sea," Science 18, no. 4092 (June 22, 1973): 1266-67.

7. J. C. Davis, "LNG: Growth or Safety," Chemical Engineering, May 28, 1973, pp. 50-52.

8. A. L. Robinson, "Energy Storage (I): Using Electricity More Efficiently," Science 184, no. 4138 (May 17, 1974): 785-87; pt. II in Science 184, no. 4139 (May 24, 1974): 884-87.

9. D. P. Gregory et al., "A Hydrogen-Energy System," American Gas Association, Alexandria, Va., August 1972.

10. R. A. Hein, "Superconductivity: Large-Scale Application," Science 185 (July 19, 1974): 211-22.

11. D. P. Snowden, "Superconductors for Power Transmission," Scientific American, April 1972, pp. 84-91.

12. A. L. Robinson, "Superconductivity: Surpassing the Hydrogen Barrier," Science 183, no. 4122 (January 25, 1974): 293-96.

13. R. L. Whitelaw, "Electric Power and Fuel Transmission by Liquid Hydrogen Superconductive Pipeline," in Hydrogen Energy, ed. T. N. Veziroglu (New York: Plenum, 1975), pp. 575-87.

14. W. D. Metz, "Helium Conservation Program: Casting It to the Winds," Science 183 (January 11, 1974): 59-63.

15. E. M. Kinderman, Stanford Research Institute, Menlo Park, Calif., personal communication, 1973.

16. "1974 Market Data—A Special Annual Market Report on the Hydrocarbon Processing Industry," Hydrocarbon Processing, August 1973.

17. H. K. Abdel-Aal and E. G. Peattie, "Hydrogen Opportunities in Saudi Arabia," in Hydrogen Energy, ed. T. N. Veziroglu (New York: Plenum, 1975), pp. 345-67.

# 10

## ENERGY END-USE ALTERNATIVES
## TO HYDROGEN

ENERGY UTILITIES TO CONSUMERS

### Gas

As natural gas reserves dwindle and production falls, gas utilities will naturally seek means to continue to serve their customers. In this sector, substitute natural gas (methane SNG) produced from coal offers the greatest competition to the possible adoption of hydrogen. There is strong incentive for gas distributors and gas utilities to turn to SNG rather than hydrogen because SNG and natural gas can simply be blended together. The SNG strategy allows the gas industry to confine all the changes to the gas production end of the gas business and to leave intact its present large investment in natural gas delivery systems.* The consumer, therefore, need not change any equipment, appliances, or habits. In contrast, conversion to hydrogen would require change of much of the distribution systems as well as consumer appliances [1]. Thus, both the consumer and the utility have a vested interest in SNG in preference to hydrogen.

Few end-use technologies appear to complement the use of hydrogen as a substitute for natural gas. One exception, however, is the flameless catalytic converter; this device would offer new opportunities for cooking and space heating [1, 2]. The increasing deployment of electric microwave ovens, however, threatens to undermine much of the use of gas cooking, and by the time hydrogen gas became available

---

*However, just as noted for hydrogen in Chapter 6, the existing network of transmission pipelines may not really link the correct locations.

in the home, there might be little attraction left in cooking on catalytic devices. The use of fuel cells to generate electricity in local substation units, or even in the home or business, could also be regarded as a complementing technology. If fuel cell technology were to advance greatly, conversion of gas utilities to hydrogen might be advanced slightly. Because of high retrofit costs, however, such use of fuel cells would almost certainly be limited to new construction.

### Electricity

Because the largest use of gas in homes and commercial establishments is for space heating (about 70 percent for residences and about 65 percent for commercial) [3], essentially any improvement in electric space-heating technology would undermine the future gas market and, thus, the prospects for hydrogen. Recent evaluations have shown that modern electric "heat pumps" can provide space heating at a cost and total energy efficiency (from the basic resource through to sensible room heat) nearly identical to that of conventional gas combustion heaters [4]. Since heat pumps can be used for cooling as well, they are likely to gain favor in new construction. Although heat pumps formally had a reputation for poor reliability, this problem is now much reduced.

Development of simple, modest-scale thermal energy storage devices (similar to those long used in some foreign countries) would enable electric utilities to encourage use of electric power for space heating further. The adoption of a dual pricing structure with lower rates for use of off-peak power (widely practiced in Europe) would enable electric utilities simultaneously to undercut the gas home-heating market and to help level their load profile. Load leveling would also reduce the need for hydrogen or other large-scale energy storage schemes in electric utilities.

Thus, several changes in the electric power sector would greatly affect the future role of hydrogen delivered to homes and commerce.

### TRANSPORTATION

### Private Automobiles

Hydrogen used to fuel private automobiles has much competition, no matter whether the hydrogen is stored as a liquid or as metal hydride. Competition comes not only from alternative fuel choices, but also from alternative automotive power plants. Some competitors even rival hydrogen's highly touted "clean air" advantage.  The leading automotive technological alternatives include [5, 6]:

Fuels
    Synthetic gasoline
    Synthetic diesel
    Methanol
    Gasoline/methanol blends
    Propane
    Liquid methane
    Onboard decomposition of fuels (or ammonia) to hydrogen [7]
    Hydrogen
Electric vehicles
    Battery [8-10]
    Flywheel [11, 12]
    Fuel cell
Engines [13]
    Gas turbines
    External combustion
    Rotary
    Modified reciprocating, such as stratified charge

Without a doubt, by the time hydrogen could be made generally available for automotive use, many of the alternatives could also be put into use.

Automotive fuel is only one part of the personal transportation system.  The complete automotive system consists of vehicles, the fuel network, sales and service, and roadways.  Consumers will favor those options that maximize the flexibility of personal mobility without undue increase in out-of-pocket costs.  Since the present automotive system already provides a high degree of personal mobility, the consumer really is faced with choosing the options that least diminish present mobility levels.

The options that cause the least interference with the normal operation of the system would be the easiest to implement.  For example, as long as it does not require a special fuel, it is fairly easy to introduce a new engine because it can be done incrementally on a vehicle-by-vehicle basis.  A new engine does not affect the complex workings of the fuel distribution network.  In contrast, introduction of an engine modification that requires a change in the fuel distribution network would prove difficult because it implies a change that must encompass the entire fuel network.  Recent history provides excellent documentation: The introduction of the exhaust catalytic converter to control air pollution proved awkward because national availability of no-lead gasoline had to be assured to both consumers and manufacturers.

The coordination problems were very great and required considerable debate [14]. Obviously, provision of no-lead gasoline entails a far less major change than a conversion from gasoline to hydrogen.

Vehicle manufacturers will certainly seek to confine change to the options that pose the least business risk. The contemplation of a change in the automotive system that requires both a basic change in the vehicle—such as provision for liquid hydrogen fuel—and also the widespread deployment of a new fuel system inevitably recalls the query whether the chicken or the egg came first in the minds of manufacturers, fuel producers, and consumers. The first sales are the most difficult to imagine. Consequently, the options that either preserve the viability of the existing fuel network or rely upon the prior existence of another widespread network (such as the electric power system) would be more easily introduced. Table 10.1 shows the compatibility of various options with relevant existing systems.

The electric car is a hydrogen competitor with properties relatively favorable to market penetration, at least for vehicles with limited range and performance. Although the electric car would not preserve the existing gasoline network, in its initial deployment both the manufacturer and the consumer could feel confident that a recharge could be obtained virtually anywhere in the country—even, perhaps, in more locations than gasoline can be purchased. The electric car would possess "clean air" attributes at the point of energy end-use comparable to the hydrogen car. Both would concentrate the associated pollution at a central fuel or power production facility.

Although it is not possible to say how much individuals and society are willing to pay to avoid drastic change, there is abundant evidence that the amount is large. As a result, the alternatives to hydrogen-fueled private cars, even the options that would themselves require strenuous exertion of the national will—such as synthetic gasoline from coal, or electric cars—appear to possess a far greater chance of being put into practice.

The key technologies complementing hydrogen use in automobiles are cryogenic liquids handling and storage, lightweight low-cost metal hydrides using abundant materials, and engine alloys that resist hydrogen environment embrittlement.

## Fleet Vehicles

Fleet vehicles, especially those that do not range far from their home bases, often do not use the public fuel distribution network; instead they often refuel only at privately owned fuel terminals. This practice stems partly from convenience but is largely the result of the reduced fuel cost made possible by bulk sales to fleet accounts. Fleet

TABLE 10.1

Compatibility of Automotive Technology Options with Energy or Fuel Distribution Networks

| Option | Existing Networks | | | Future Network |
| | Gasoline/Diesel | Natural Gas | Electric | $H_2$ Gas Utility |
|---|---|---|---|---|
| **Fuels** | | | | |
| Hydrogen | | | | |
| Liquid | P | P | — | F |
| Gaseous | P | F | — | E |
| Metal hydride | P | P | — | E |
| Onboard decomposition | E | F | — | — |
| Methane | | | | |
| Liquid | P | F | — | — |
| Gaseous | P | E | — | — |
| Methanol | G | — | — | — |
| Ethanol | G | — | — | — |
| **Prime mover** | | | | |
| Battery/electric | — | — | E | — |
| Fuel cell | F | G | — | G |
| Turbine | E | G | — | G |
| External combustion | E | G | — | G |

Note: E = excellent, G = good, F = fair, P = poor, — = not applicable.

vehicles are much more amenable to hydrogen fueling because a single fueling point can often serve the entire fleet. Thus, the operator of the fleet, having a focused decision-making capability, can mandate the fuel used in the many vehicles of the fleet in a single decision. This single decision stands as an important contrast to use of the private automobile, in which each consumer makes an independent decision whether to purchase a vehicle using an unconventional fuel.

Naturally, however, the fleet operator must still be concerned with the basic availability of an energy source to propel his vehicles. Since fleet owners usually employ business accounting procedures, they can be expected to place more emphasis on the relative cost of options and less emphasis on convenience than the private automobile owner.

Fleet operators of small vehicles, especially those whose vehicles have limited need for long-distance excursions, can be expected to view electric vehicles as strong competitors to hydrogen fuel. In some foreign countries, small fleet vehicles have long been electric (English milk delivery vans, for example), and the U.S. Post Office has recently made a commitment to a small test fleet of battery-electric mail delivery vans [15]. There is under way a demonstration experiment of a bus powered by an advanced flywheel that is reenergized electrically [12]. Thus, considerable momentum is already gathering behind the electric-powered fleet vehicle option [16].

Fleets that operate over long distances, however, such as intercity trucks and buses, are generally poor candidates for electric propulsion. Provided that they operate only over relatively fixed routes between repetitive end-points, these vehicles could easily shift to an unconventional fuel such as hydrogen, methanol, liquid methane, or propane. Indeed, many fleet trucks and buses already operate on diesel—a fuel that is unconventional for automobiles.

Although fleet vehicles are more likely to be able to adopt hydrogen than private automobiles, the alternatives pose stiff competition and there is already momentum behind several alternatives.


Ships and Trains

Propulsion of ships with hydrogen faces competition from nearly every other fuel, natural or synthetic, except electricity. Although nuclear power has long been an option, it has been adopted only by the military, where long range and staying power weigh more heavily than operating cost. The nuclear-powered demonstration freighter Savannah, a highly touted example of the Atoms-for-Peace program of the 1950s, has not been imitated.

Trains can also be operated with nearly any fuel or with electric power. Electric power is a strong contender, while direct use of nuclear power is a very weak alternative. For turbine trains, hydrogen is probably an effective fuel but is not as easily handled as simpler liquids such as methanol. For magnetic levitation trains, no fuel other than hydrogen offers the same hope for use as both fuel and refrigerant for superconducting magnets.

## Commercial Aircraft

Liquid hydrogen is the most viable unconventional aviation fuel by a wide margin, principally because of its unrivaled energy density per unit weight [17-18]. Hydrogen's long-term major competitor is synthetic jet fuel, chemically similar to present fuels, but derived from coal or oil shale rather than natural petroleum. Liquid methane contains appreciably less energy per unit weight than liquid hydrogen yet has comparable cryogenic handling and containment problems. Consequently, a conversion to liquid methane would entail nearly the same difficulties as hydrogen, without comparable benefits.

Methanol contains only about half the energy per unit weight as conventional jet fuel. Consequently, to maintain a given range, the weight of methanol carried by the airplane would have to be roughly twice that of conventional fuels; this would adversely affect the net energy efficiency and the revenue-producing payload of the flight. The greatest advantage of methanol compared with hydrogen is that it could be handled by existing ground equipment, because it is a liquid at ambient temperature. Although methanol would offer an easier transition, it would greatly impair the physical and economic performance of the aircraft and is, therefore, an unlikely candidate.

## IRON ORE REDUCTION

Numerous chemicals can be used as reducing agents, but probably only hydrogen and carbon compounds can be made available in the large quantities needed. Besides the coke presently used, charcoal from any source (such as wood-processing waste) or carbon monoxide might be used. Only hydrogen seems to be a viable alternative to conventional coking coal methods that does not depend on a fossil-fuel resource. Any technology that lowers the cost of hydrogen production would complement this application.

## SUMMARY

In nearly every end-use sector, as long as fuels derived from fossil resources are available, hydrogen faces strong competition from alternative technological options. The two sectors in which hydrogen appears to be the superior option are commercial aircraft fuel and iron ore reduction.

## REFERENCES

1. D. P. Gregory et al., "A Hydrogen-Energy System," American Gas Association, Alexandria, Va., August 1972.

2. J. C. Sharer and J. B. Pangobrn, "Utilization of Hydrogen as an Applicance Fuel," in Hydrogen Energy, ed. T. N. Veziroglu (New York: Plenum, 1975), pp. 875-87.

3. "Patterns of Energy Consumption in the United States," Stanford Research Institute for the Office of Science and Technology, Washington, D.C., January 1972.

4. "Does Electric Heating Waste Energy?" Electrical World, May 1, 1973, p. 66.

5. F. H. Kant et al., "Feasibility Study of Alternative Fuels for Automotive Transportation," vol. I: Executive Summary, vol. II: Technical Section, vol. III: Appendices, Exxon Research and Engineering Company, Linden, N.J., for the U.S. Environmental Protection Agency, June 1974.

6. J. Pangborn and J. Gillis, "Alternative Fuels for Automotive Transportation—A Feasibility Study," vol. I: Executive Summary, vol. II: Technical Section, vol. III: Appendices, Institute of Gas Technology, Chicago, Ill., for the U.S. Environmental Protection Agency, June 1974.

7. "Hydrogen Fueled Car Runs Pollution Free," Industrial Research, November 1973, pp. 25-26.

8. P. A. Nelson et al., "The Need for Development of High Energy Batteries for Electric Automobiles," Argonne National Laboratory Report ANL 8075, November 1974.

9. R. Connolly, "Electric Vehicles Draw Interest," Electronics, March 7, 1974, pp. 70-72.

10. "Switching on Electric Vehicles," Environmental Science and Technology, May 1974, pp. 410-11.

11. R. F. Post and S. F. Post, "Flywheels," Scientific American, December 1973, pp. 17-23.

12. "A Spinning Flywheel Provides the Power," Business Week (Industrial Edition), August 11, 1973, pp. 84-D-84K.

13. D. G. Harvey and W. R. Menchen, "A Technology Assessment of the Transition to Advanced Automotive Propulsion Systems,"

Hittman Associates, Inc., Columbia, Md., for the National Science Foundation, May 1974.

14. "Gasoline: EPA Issues Rules, Proposal on Availability of Unleaded Gasoline," Energy Users Report, May 9, 1974, pp. C-2, C-3.

15. "When Electric Trucks Deliver the Mail," Business Week (Industrial Edition), March 9, 1974, pp. 78C-78F.

16. "Now Otis Moves Horizontally," Business Week (Industrial Edition), March 9, 1974, pp. 78J-78M.

17. "Research and Development Opportunities for Improved Transportation Energy Usage," report of the Transportation Energy Panel, PB220612, National Technical Information Service, Springfield, Va., September 1972.

18. G. D. Brewer, "The Case for Hydrogen-Fueled Transport Aircraft," Astronautics and Aeronautics, May 1974, pp. 40-51.

# ECONOMICS OF HYDROGEN

The place of hydrogen in the energy economy of the future will depend heavily on its ability to compete with other fuels in the marketplace. Consumers will be willing to pay only a price that is consistent with their own evaluation of the comparative advantages of hydrogen. This chapter presents the costs of delivering hydrogen to the market for each important step—production, distribution, liquefaction, and storage. The economic relationships between hydrogen and other fuels are also explored briefly.

# 11

## HYDROGEN COSTS AND ECONOMIC
## RELATIONSHIPS TO OTHER FUELS

### DEVELOPMENT OF COST ESTIMATES

Recent literature [1, 2, 3, 4] has been used to establish estimates for the cost of the various hydrogen technologies. In literature pertaining to a hydrogen economy there is a wide difference of opinion about future costs of hydrogen in all stages of the production-consumption cycle. These differences stem from varying assumptions about capital investment, financing methods (utility versus industrial), interest rates, rate of return on investment, depreciation rates, and overhead expenses. Although such discrepancies are normally expected, it is difficult for readers to reconcile the various estimates unless the assumptions are spelled out. To provide a consistent set of costs, we have developed cost estimates using uniform assumptions, whenever possible, for capital investment, annual fixed charges, operating costs, and energy costs.

Unfortunately, the changing value of the dollar complicates the interpretation and reconciliation of cost estimates. The recent rapid rate of inflation in the United States—especially in the construction sector, in which costs were escalating more rapidly than in other sectors—makes it necessary to determine the dollar-year on which the estimate is based and then to adjust the estimate. In this book, all cost estimates have been adjusted to mid-1975 dollars by means of the indexes given in Figure 11.1. This adjustment procedure, unfortunately, adds uncertainty to the estimates quoted.

In this book, annual fixed charges are calculated from the capital investment by using an annual factor of 0.20. The components of this annual factor are shown below:

## FIGURE 11.1

### Comparison of Price Indexes

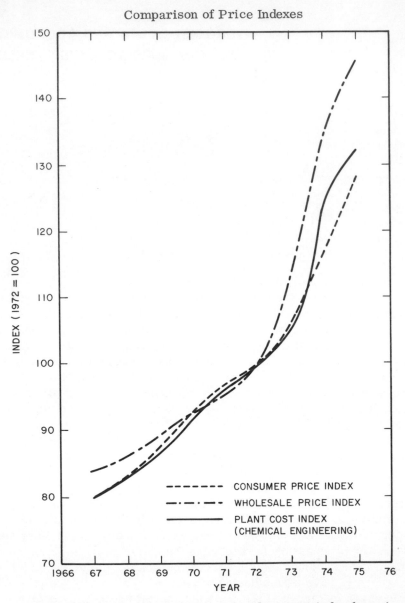

Note that in the last few years the plant cost index has risen more sharply than the other indexes.

Sources: Survey of Current Business, various issues, Department of Commerce, Bureau of Economic Analysis; Chemical Engineering, various issues.

1. Capital charges                                          0.13
   a. Interest rate on debt @ 9 percent
   b. Rate of return on equity @ 15 percent
   c. Equity financing @ 65 percent
2. Depreciation, assuming a 20-year life for
   tax purposes                                             0.05
3. General overhead, taxes, and insurance                  0.02
                                                            ‾‾‾‾
                                                            0.20

The assumption of a 9 percent interest rate and 15 percent return on
equity is in line with long-term trends for these parameters.* Although the prime interest rate has exceeded 9 percent in periods of
tight money, it is quite likely that over the long run the federal government will endeavor to keep interest rates below 10 percent. The ratio
of debt to equity financing in U.S. corporations has been increasing in
recent years owing to conditions in the stock market that have made
it difficult for corporations to raise equity capital. For manufacturing corporations, debt composed 30 percent of total financing in 1973
versus 20 percent in 1964 [5]. New financing has been running 80 to
90 percent debt, and for many corporations debt exceeds equity, even
though 65 percent equity and 35 percent debt is usually considered
desirable. The weighted cost of capital shown above was based on
35 percent debt financing and 65 percent equity; the capital charge
factor, then, is calculated as follows: $[0.09 \times 0.35] + [0.15 \times 0.65]$
$= 0.13$.

By including an assumed rate of return on equity in the fixed
charge factor, the production costs presented in this report represent
the total revenue required not only to produce the good, but also to
reap an adequate return on investment. In many aspects of the hydrogen economy, the cost of energy represents a significant part of the
total cost estimate; consequently this is shown separately from other
operating costs.


## HYDROGEN PRODUCTION COSTS

As emphasized earlier, the primary hydrogen resources are
fossil fuels and water. Today most hydrogen is produced from either
natural gas or naphtha; electrolysis of water is the only other commercially significant source, but it remains minor compared with the
fossil-fuel sources. The economics of several production methods,

―――――――――――――――

*These are the same values found in the Project Independence
Report (reference 6).

those that have been proved technically and commercially as well as some that are still experimental, are presented below.  Hydrogen from natural gas was omitted because it cannot be a long-term source in the United States, given current shortages and rising prices.


## Hydrogen from Coal

The availability of coal to produce hydrogen will depend on competing demands for coal as well as on federal and state regulations concerning both underground and surface mining.  There are various estimates of total U.S. coal reserves, but one of the most respected is shown in Table 11.1.  At current rates of consumption, there is optimism regarding the longevity of coal reserves.  However, the possible switch from petroleum fuels to coal by several large energy-using sectors of the economy would considerably shorten the longevity of the coal reserves.

The costs of hydrogen produced from gasification of coal presented here are based on current steam oxygen technology.  The particular technology considered is a low-pressure, high-temperature gasifier licensed by Koppers-Totzek.  For costing purposes a capacity of 318 million standard cubic feet (SCF) of hydrogen per day has been assumed.  Coal consumption for both gasification and process heat is assumed to be 7,500 tons per day, which is equivalent to 0.085 ton of coal per $10^6$ Btu of hydrogen output.


## TABLE 11.1

U.S. Coal Reserves* and Recent Production

| Category | Quantity (billions of tons) |
|---|---|
| Demonstrated, in place | 434.0 |
| Accessible to surface mining | 137.0 |
| In lower 48 states | 129.0 |
| Recoverable (at 80 percent recovery) | 103.0 |
| Production (1974) | 0.6 |

*"Reserves" measure the amount of "resource" recoverable with present technology at current price levels.  When either price levels or technology changes, the amount of resource designated as reserves also changes.

Source: Reference 7.

## TABLE 11.2

### Cost of Producing Hydrogen from Coal*
### (1975 dollars)

|  | $/10^3$ SCF | $/10^6$ Btu |
|---|---|---|
| Capital cost | 0.61 | 2.22 |
| Operating cost | 0.13 | 0.48 |
| Coal feed @ $15/ton | 0.34 | 1.27 |
| Total | 1.03 | 3.97 |

*See text for assumptions.

The 1975 estimated capital cost of a plant of this size is $320 million, including working capital and the required ancillary oxygen plant. The investment cost works out to $1.00 per SCF of hydrogen per day. Operating costs, including purchased electricity at 1¢/kWh, are estimated to be $13.8 million per year. Although these costs assume gasification takes place at the mine mouth, mine investment is not explicitly included but is covered in the cost of coal. Coal is assumed to have a heating value of 24 million Btu per ton, which is typical of eastern coal (western coal is typically 18 million Btu per ton). Assuming annual production is $105 \times 10^9$ SCF/yr (over a 330-day year), the unit operating cost is $0.13/10^3$ SCF or $0.48/10^6$ Btu of hydrogen. Annual capital costs were estimated by multiplying the capital factor of 0.20 (noted above) by the total capital investment.

The costs of producing hydrogen from coal by the Koppers-Totzek process operated at the mine mouth are summarized in Table 11.2 for eastern coal costing $15 per ton.* Figure 11.2 shows the variation in current production costs as a function of the cost of coal. Capital and operating costs that do not vary with coal costs total $2.70 per $10^6$ Btu of hydrogen output. Technical advances should lower production costs in the future (see Table 11.4).

---

*AMAX Inc., a major coal producer, has stated the following revenue requirements to yield a 15 percent discounted cash flow return on investment in 1975 dollars (reference 18): eastern underground, $16.00/ton; eastern surface, $13.00/ton; western surface, $5.00/ton.

FIGURE 11.2

Cost of Hydrogen from Coal as a Function of Coal Cost

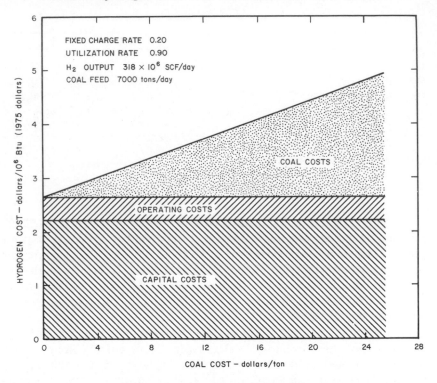

### Hydrogen from Electrolysis of Water

The production of hydrogen by electrolysis of water has long been commercially feasible, and electrolytic hydrogen plants have been operating for over 40 years [1]. The largest is at Rjakon, Norway, and has a capacity of 284,000 lbs of hydrogen per day. Technology used in existing plants is obsolescent. The long-term economic feasibility of electrolysis will depend on low capital costs, high efficiency, and, most important, the cost of electricity. The production cost of hydrogen can also be credited with by-product sales of oxygen whenever a market exists. Currently, however, electrolytic hydrogen is expensive.

The costs of electrolytic hydrogen presented here are estimated for an advanced technology proposed by Allis-Chalmers in 1966 (and further developed by Telednye Isotopes) that is not yet in commercial use and are adapted from the report of the Synthetic Fuels Panel [1]. The data presented by the Synthetic Fuels Panel were optimized for operation with electric power costing up to $0.008/kWh; however, even though electric costs will soon be $0.01/kWh or more, we have made no attempt to reoptimize component engineering.

Investment is estimated at $53 per lb of hydrogen per day. Assuming 90 percent utilization and fixed capital charges of 20 percent per year, the capital costs work out to $0.03/lb of hydrogen or $0.63/10^6$ Btu. Operating costs were estimated at $0.014/lb of hydrogen or $0.27/10^6$ Btu. Electrical energy consumption was given as 20 kWh/lb of hydrogen [1]. Because of the vast amounts of electricity consumed in the process, the cost of electrolytic hydrogen depends heavily on the cost of electricity. Figure 11.3 shows the total cost of electrolytic hydrogen as a function of electricity cost; for purposes of comparison, the costs of currently available technology are also shown. The current technology case consumes 24 kWh/lb of hydrogen and has a unit investment cost about 240 percent that of the advanced technology. Production costs, excluding electricity, for current and advanced technology are presented in Table 11.3.

## Hydrogen from Thermochemical Processes

The use of nuclear heat as a source of energy for the thermochemical decomposition of water is an attractive prospective technol-

### FIGURE 11.3

Estimated Production Costs for Electrolytic Hydrogen as a Function of the Cost of Electricity

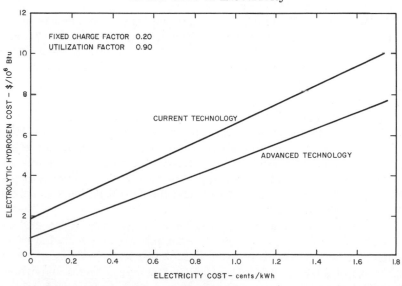

Source: Adapted from reference 1.

TABLE 11.3

Production Costs of Electrolytic Hydrogen Excluding Electricity Costs
(1975 dollars)

|  | Dollars per $10^6$ Btu | |
|---|---|---|
|  | Present Technology | Advanced Technology |
| Capital charges | 1.50 | 0.63 |
| Operating costs | 0.36 | 0.27 |
| Total | 1.86 | 0.90 |

Source: Adapted from reference 1.

ogy for the production of hydrogen because it offers potentially superior energy efficiency. Thermochemical decomposition is expected to have an economic advantage over the electrolytic production process because the intermediate step of converting nuclear heat to electrical energy is eliminated, thereby reducing the need for expensive generating equipment and, more important, the attendant energy losses. However, the cost of thermochemical hydrogen production facilities is quite uncertain because the possible processes have not advanced beyond laboratory evaluation.

The Futures Group [3] analyzed the Mark I thermochemical process as described by deBeni and Marchetti of Euratom [7] and estimated the capital cost of a plant producing 2 million SCF of hydrogen per hour at $74 million (1972 dollars) exclusive of nuclear reactor costs. In 1975 dollars this would become $98 million. Fixed charges and capital costs amounted to $4.50 per $10^6$ Btu; adding operating costs of $3.25 per $10^6$ Btu and a nominal charge of $0.33 per $10^6$ Btu for nuclear heat yields a total hydrogen cost of $8.08 per $10^6$ Btu.

Chao and Cox [8] presented some admittedly speculative estimates based on 1972 data. Their "pessimistic case" of estimated total capital costs for a nuclear reactor and chemical plant works out at $232 per kW (thermal) in 1975 dollars; the resulting hydrogen cost is $3.11 per $10^6$ Btu.

Wentorf and Hanneman [4] of General Electric estimated capital costs for nuclear heat from high-temperature gas-cooled reactors. Adjusted, their estimate is $160 to $265 per kW. Assuming nuclear fuel costs of $0.32 per $10^6$ Btu, utility financing, and a 90 percent utilization rate, the adjusted cost of input nuclear heat was estimated at $1.32 to $1.96 per $10^6$ Btu. These authors speculate that thermochemical processes for commercial use will have a net thermal efficiency of 50 percent.  Therefore, the estimated nuclear heat costs for hydrogen will be in the range of $2.64 to $3.92 per $10^6$ Btu in 1975 dollars.  In this reference the capital and operating costs for the hydrogen generation equipment are based on "an approximate analysis" of some candidate thermochemical processes, and are estimated to add at least $1.30 per $10^6$ Btu.  Thus, they estimate that the total adjusted cost of hydrogen will be in the range of $3.90 to $5.25 per $10^6$ Btu. *

Since none of the references discussed above has included a complete discussion of the capital cost factors used to derive the cost of hydrogen, it is impossible to assess the realism of their economic estimates.  Clearly, however, the eventual cost of thermochemical hydrogen depends primarily on the cost of the nuclear heat source and the cost of thermochemical conversion processes; the latter is especially uncertain at this time.  Moreover, it is difficult to extract cost estimates for nuclear-produced heat from published costs of nuclear-produced electricity because estimates for nuclear power installations include electrical generation equipment.  Published cost estimates are also difficult to reconcile because it is often unclear whether allowance is made for the cost of funds invested during construction and cost escalation of equipment, materials, and labor.  These factors are especially important, because over the eight- to ten-year construction period for a light water reaction, these two expense categories can add up to 50 percent of the total capital costs.  A more extensive discussion of problems encountered in estimating costs of nuclear plants is presented in Chapter 16.

Highly focused solar energy is another heat source that has been proposed for thermochemical decomposition.  However, precision, steerable, focused solar collectors are so expensive that a solar-powered system is unlikely to be economically competitive with nuclear heat for a long time.  Nevertheless, very optimistic and almost certainly unrealistic estimates [9] give hydrogen costs of approximately $5.32 per $10^6$ Btu in 1985, decreasing to $3.66 per $10^6$ Btu in 2000.

---

*Published comment on Wentorf and Hanneman's paper is highly critical of the lowness of their cost estimates.  (See, for example, reference 19.)  Since their estimates were published, General Electric announced that it was shelving investigation of thermochemical hydrogen production methods.  This move could be interpreted as verification that their estimates were unrealistically low.

Summary Comparison of Hydrogen Production Costs

Cost estimates for the production of hydrogen from coal, electrolysis of water, and thermal decomposition of water prepared in
this study and by others are compared in Tables 11.4 through 11.6.
Although we have made adjustments in energy, capital, and operating
costs whenever possible, some discrepancies still exist because of
unstated assumptions in the various sources. In general, the capital
investment costs and process efficiencies are the most difficult to alter responsibly, but we have used the plant cost indexes published in
Chemical Engineering to adjust the capital costs to 1975 dollars. The
resulting cost comparisons are presented to indicate the diversity of
technical and economic opinion that makes the estimation of hydrogen
production costs difficult for the long term. It is clear, however,
that production of hydrogen depends on other energy sources and will,
therefore, be more costly than basic fossil and nuclear energy sources.
Any net economic advantages of hydrogen must come from end-use
system efficiency gains rather than from production costs.

The estimates in the tables range from a low of \$1.60 per $10^6$
Btu to a high of \$7.96 per $10^6$ Btu. The lowest figures are for hydro-
gen made from coal, which is a well-known technology with many of
the key components of equipment currently in use. The Linde estimate is considerably lower than this study's because of a lower esti-

TABLE 11.4

Cost Estimates for Producing Hydrogen from Coal
(1975 dollars)

|  | This Study[a] | Linde Division, Union Carbide[b] |
|---|---|---|
| Year of original estimate | 1974 | 1973 |
| Dollar year | 1975 | 1973 |
| Energy source | Coal | Coal |
| Cost per unit | \$15/ton | \$15/ton |
| Plant capacity | $87 \times 10^9$ Btu/day | $110 \times 10^9$ Btu/day |
| Unit investment | \$3,640/$10^6$ Btu/day | \$1,440/$10^6$ Btu/day |
| Unit operating cost | \$0.48/$10^6$ Btu | \$0.13/$10^6$ Btu |
| Operating factor | 0.90 | 0.90 |
| Gaseous hydrogen cost | \$3.97/$10^6$ Btu | \$2.08/$10^6$ Btu |

[a]See Table 11.2.
[b]Hypothesized advanced technology, adapted from reference 2.

mate of capital costs resulting from current research and development [2].

For future electrolysis plants, hydrogen production costs shown in Table 11.5 cluster in the $3.80 to $4.80 per $10^6$ Btu range. Diverse assumptions about capital costs, financing, depreciation, and electrical consumption rates underlie most of the differences among estimates. Since the basic technology is the same in most of the references cited, namely, an advanced cell proposed by Allis-Chalmers, it is reasonable to expect some agreement among the various cost estimates.

Thermochemical decomposition of water is by far the most speculative of the hydrogen production methods. Although the cost estimates in Table 11.6 are the best available, they vary by more than a factor of two even though each estimate is based on technology developed at Euratom [1]. The past history of nuclear power suggests that cost escalation (even in constant dollars) is likely to be significant in the next decade, and thus the adjusted Futures Group [3] estimate of $7.96 per $10^6$ Btu appears much more valid than estimates in the range of $2 to $4 per $10^6$ Btu.

## HYDROGEN TRANSPORT COSTS

The cost of transmitting hydrogen will depend on the mode of transportation and the form of the hydrogen, liquid or gaseous. Current hydrogen deliveries in the United States are made by cryogenic tank trucks and railcars; while relatively expensive, these offer a flexible mode for the transition period. The following discussion will emphasize the costs of distributing gaseous hydrogen in pipelines, because this appears to be the sole feasible method of transmitting vast amounts of hydrogen.

### Gaseous Hydrogen

Until recently, analyses of the cost of hydrogen transmission were all based on the adaptation of natural gas pipelines to the transport of hydrogen [2, 3, 11-13]. The newer, more credible, estimates take into consideration a pipeline design optimized expressly for hydrogen based on the difference in the physical properties of hydrogen and natural gas. The most complete of these recent studies is that conducted by the Institute of Gas Technology, in which transmission costs for a variety of pipeline diameters and operating pressures are calculated [12]. Based on that study, the 1975 cost of gaseous hydrogen transmission is estimated to lie between $0.033 and $0.09 per million Btu for 100 miles, depending on pipe diameter,

TABLE 11.5

Adjusted Cost Estimates for Producing Hydrogen by Electrolysis of Water
(1975 dollars)

| | Synthetic Fuels Panel (reference 1) | | General Electric (reference 10) | Institute of Gas Technology (reference 11) | Linde Division of Union Carbide (reference 2) |
|---|---|---|---|---|---|
| | Current Technology | Advanced Technology | | | |
| Time frame for original estimate | 1974 | 1974 | 1990 | 1972 | 1973 |
| Dollar year | 1972 | 1972 | 1973 | 1972 | 1973 |
| Electricity | | | | | |
| Unit cost (cents per kWh) | 1.0 | 1.0 | 1.0 | 1.0 | 1.0 |
| Consumption (kWh per lb) | 24.0 | 20.0 | 17.7 | 18.2 | 21.7 |
| Unit investment cost (dollars per $10^6$ Btu per day) | 2,440 | 1,030 | 725 | 1,415 | 1,010 |
| Unit operating cost (dollars per $10^6$ Btu) | 0.36 | 0.27 | — | — | 0.09 |
| Operating factor | 0.90 | 0.90 | 0.85 | 0.95 | 0.88 |
| Fixed cost factor | 0.20 | 0.20 | 0.15 | 0.21 | 0.17 |
| Gaseous hydrogen cost (per $10^6$ Btu) | 6.52 | 4.78 | 3.78 | 4.39 | 4.83 |

TABLE 11.6

Adjusted Cost Estimates for Producing Hydrogen by Thermochemical Decomposition of Water (1975 dollars)

| | Chao and Cox (reference 8) | Wentorf and Hanneman (reference 4) | Futures Group (reference 3) |
|---|---|---|---|
| Year of original estimate | 1974 | 1974 | 1972 |
| Dollar year | 1972 | 1972 | 1972 |
| Plant capacity | | | 2 million SCF per hour |
| Unit investment cost[a] | ($132–232) per kW (th) | ($160–265) per kW (th) | $7,400 per $10^6$ Btu per day |
| Unit operating cost | | $1.30 per $10^6$ Btu[b] | $2.00 per $10^6$ Btu |
| Operating factor | | 0.90 | |
| Gaseous hydrogen cost (per $10^6$ Btu) | $2.00–3.11 | $3.90–5.25 | $7.96 |

[a]Chao and Cox include nuclear and chemical plant; Wentorf and Hanneman present cost for nuclear plant only; the Futures Group presents the chemical plant only.
[b]Includes capital cost.

operating pressure, and hydrogen costs. These estimates apply to a
hydrogen-optimized system operating at 750 psia. Transmission
costs vary with the cost of hydrogen because hydrogen is drawn from
the pipeline to fuel the pipeline compressors. The sensitivity of pipeline transmission costs to compressor spacing appears to be slight
for a 1,000-mile pipeline. This is shown in Table 11.7, in which hydrogen transmission costs are given for pipelines operating at 750
psi and 1,500 psi for three different costs of hydrogen. The figures
in Table 11.7 were adjusted to reflect a 20 percent annual fixed charge
rather than the 13 percent charge assumed in reference 12. The original data are roughly 8 percent lower. This adjustment was based on
the relationship between fixed and operating costs in the related study
cited earlier [11].

## Liquid Hydrogen

### Liquefaction

The major steps in liquefaction of hydrogen are purification,
refrigeration, and conversion-liquefaction [14]. Two estimates of
the costs involved in this process are treated here, one by the Linde
Division of Union Carbide Corporation published in 1973 [2] and one
by Air Products and Chemicals, Inc., published in 1968 [15]. Both
estimates are based on a plant producing 250 tons of liquid hydrogen
per day.

TABLE 11.7

Cost of Gaseous Hydrogen Transmission in Optimized Pipelines for
Various Assumed Costs of Hydrogen Used to Fuel Compressors
($\text{¢}/10^6$ Btu per 100 miles; 1975 dollars)

| Pipeline Diameter (in) | Operating Pressure (psia) | Cost of Hydrogen for Pipeline Compressors | | |
|---|---|---|---|---|
| | | $2.00/10^6$ Btu | $3.00/10^6$ Btu | $4.00/10^6$ Btu |
| 24 | 750 | 6.3 | 7.3 | 8.8 |
| | 1,500 | 5.0 | 5.9 | 6.8 |
| 36 | 750 | 4.1 | 4.7 | 5.1 |
| | 1,500 | 3.6 | 4.0 | 4.5 |
| 48 | 750 | 3.7 | 4.1 | 4.6 |
| | 1,500 | 3.5 | 3.9 | 4.4 |

Source: Adapted from reference 12.

The adjusted Linde estimated investment is $28.4 million for liquefaction equipment for a plant of 250-ton capacity. Assuming financing typical of refining and chemical companies, a fixed charge factor of 17 percent per year was applied, composed of 5 percent for depreciation (15 years), 8 percent for average interest and return on investment, and 2 percent for taxes and insurance. An additional 3 percent was added to cover operating maintenance and labor. Assuming an operating time of 320 days per year (88 percent) and an average throughput of just under $1,200 \times 10^6$ Btu/hr, the resulting fixed charges amounted to $0.62 per $10^6$ Btu. Using nuclear electric energy at $0.01 per kWh yields a cost of purchased energy at $1.06 per $10^6$ Btu. Thus, the total liquefaction cost was $1.68 per $10^6$ Btu. This implies roughly 5 kWh per pound of liquid hydrogen.

When the cost estimates in the 1968 study by Air Products [15] are adjusted to 1975 dollars, capital investment and operating costs are increased by 60 percent. Capital investment required becomes $50 million for a $250 ton per day plant, while annual operating costs become $3 million. The annual costs become $10 million for fixed charges plus $3 million for operating costs (or $0.079 per lb). Electricity consumption was specified at 4.46 kWh per lb of hydrogen [15]. Assuming $0.01 kWh for electricity costs, the total cost of hydrogen liquefaction is $0.124 per lb ($0.079 + $0.0446). This yields hydrogen at $2.40 per $10^6$ Btu—roughly $0.72 per $10^6$ Btu higher than the Linde estimate. The difference stems largely from the lower unit investment used by Linde. Figure 11.4 shows the two cost estimates as a function of electricity prices; the two lines converge because the Air Products study assumed a slightly lower electricity consumption rate than that in the Linde estimate.

## Liquid Transport

Transport of liquid hydrogen is more difficult and therefore more costly than the transport of petroleum fuels because of the necessity of cryogenic facilities.

From another Linde report [16] the adjusted cost of transporting liquid hydrogen 100 miles by truck becomes $0.50 per $10^6$ Btu, or roughly six to ten times the cost of gaseous hydrogen transport by pipeline. The adjusted cost of barge transport for 100 miles is estimated at $0.095/10^6$ Btu, which is comparable to the cost of gaseous hydrogen carried in a 24-inch pipeline [16]. Rail transport of liquid hydrogen for 100 miles has been estimated (adjusted) to cost $0.60/10^6$ Btu, based on a 500-mile round trip, a freight rate of one cent per ton mile, and amortization of the capital costs of tank cars at 12 percent per year [9]. On a large scale, economies of scale would probably reduce the rail costs to make rail and truck transport roughly equal in cost—a conclusion also reached by the Futures Group [3].

FIGURE 11.4

Liquefaction Costs as a Function of Electricity Cost
(1975 dollars)

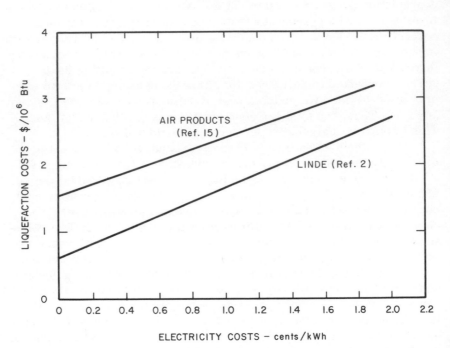

HYDROGEN STORAGE COSTS

Gaseous Hydrogen

The easiest method of storing gaseous hydrogen is to increase
pipeline pressure to provide a temporary excess inventory of gas.
This process is called "pipeline packing" and can be used only for
short periods. Nevertheless, it is attractive because the only cost is
the incremental investment in sturdier compressors and pipe.

Storing gaseous hydrogen in underground rock strata or caverns
is also inexpensive. The major problems of underground storage are
location and withdrawal rate. The location of appropriate geological
formations often is not compatible with demand centers and pipeline

routes. Withdrawal rates that are too rapid cause a pressure drop around the well. Reportedly, 1 percent of the storage volume can be removed per day if adequate gas pressure is to be maintained [3].

Gaseous hydrogen can be stored in tanks at a wide range of pressures (60 to 2,000 psi) and volumes. If the storage pressure is greater than that of the delivered hydrogen, the cost of additional compression capacity must be added to the storage costs. Table 11.8 summarizes the (adjusted) comparative investment costs for bulk hydrogen storage estimated by the Futures Group [3]. Since all of these storage methods use technologies currently in operation, these estimates are reasonably reliable.

### Liquid Hydrogen

Liquid hydrogen has been stored in the U.S. space program in single-container quantities approaching 1 million gallons. The adjusted unit investment of such large-volume storage capacity has been cited by Linde to be about $1.30 per gallon or $34 per million Btu [16]. Another (adjusted) estimate [9] places the investment at $28 per million Btu; however, it was assumed that relatively high boil-off rates were acceptable. In such cases, insulating requirements are not stringent. However, if the hydrogen is to be stored for long periods or is to be distributed later in liquid form, boil-off has to be minimized through more expensive insulation, which is presumably reflected in the Linde estimate. Operating costs are difficult to estimate because they depend on hydrogen withdrawal and refill patterns. Boil-off rates for a tank only 25 percent full are higher than for a tank 90 percent full; consequently, operating costs are higher when the average inventory is considerably less than the maximum capacity.

### TABLE 11.8

Comparative Investment Costs of Gaseous Hydrogen Storage
(1975 dollars)

| | |
|---|---|
| Low-pressure | $3,945/10^6$ Btu |
| High-pressure (cylinders) | 1,562 |
| Pipeline packing | 305 |
| Underground | 20 |

Source: Adapted from reference 3.

## COSTS OF OTHER FUELS COMPARED
## TO HYDROGEN SYSTEMS

The future costs of delivered hydrogen cannot be completely estimated because costs of storage and distribution are difficult to estimate without design of a detailed system complete with demand levels. However, some initial comparisons can be made using those system elements for which adequate cost estimates exist, namely, production, pipeline transmission, and liquefaction. Table 11.9 presents delivered costs for two simple hydrogen systems. The first system supplies gaseous hydrogen and is compatible with both residential and industrial end-uses. The second set of costs adds liquefaction costs to the first; this system is intended to indicate the cost of hydrogen when used as a fuel for automobiles. In both systems the costs of storage and distribution to end-users was omitted. The table shows that the best price consumers could hope for is \$1.76 per $10^6$ Btu for gaseous and \$3.44 per $10^6$ Btu for liquid hydrogen.

As shown in Table 11.10 the current costs of fossil fuels in 1972 dollars are less than the delivered cost of hydrogen with the sole exception of regular gasoline. However, as an automobile fuel, gasoline at \$3.96 per $10^6$ Btu approaches the cost of hydrogen from coal; however, electrolytically produced hydrogen costs far more, even with technological advances in electrolysis. To add further perspective, residential electricity at 3 cents per kWh is equivalent to \$8.79 per $10^6$ Btu.

Concerning the transition to a hydrogen economy, the key problem is the future outlook for relative energy prices. Table 11.10 compares U.S. energy prices in (then) contemporary and constant 1972

TABLE 11.9

Cost Comparisons for Representative Hydrogen Systems
(1975 dollars)

|  | From Coal @ \$10/ton | Electrolytic @ 1¢/kWh |
|---|---|---|
| Gaseous $H_2$ delivered 500 miles by gas pipeline |  |  |
| Current technology | $3.75/10^6$ Btu | $6.77/10^6$ Btu |
| Advanced technology | 1.76 | 5.00 |
| Gaseous $H_2$ delivered with liquefaction after delivery |  |  |
| Current technology | 5.43 | 8.45 |
| Advanced technology | 3.44 | 6.68 |

TABLE 11.10

Comparative Costs of Petroleum Products, 1973-74
(dollars per $10^6$ Btu)

| | Current Dollars | | 1975 Dollars | |
| | Aug. | Aug. | Aug. | Aug. |
| Customer and Product | 1973 | 1974 | 1973 | 1974 |
|---|---|---|---|---|
| Utilities | | | | |
| Coal | 0.40 | 0.77 | 0.72 | 0.83 |
| Oil | 0.77 | 1.96 | 1.49 | 2.02 |
| Gas | 0.34 | 0.52 | 0.57 | 0.60 |
| Residential | | | | |
| Gas (West) | 1.01 | 1.19 | 1.36 | 1.38 |
| Untaxed "regular" gasoline | 2.14 | 3.47 | 2.58 | 3.96 |
| ($/gal) | (0.27) | (0.43) | (0.32) | (0.49) |

Sources: Oil and Gas Journal, various issues; reference 17;
and personal communication from Pacific Gas and Electric Co.

dollars for August 1973 (before the Middle East War) and in August 1974 after the increases that resulted from new price levels imposed on petroleum by the Organization of Petroleum Exporting Countries (OPEC). Though OPEC controls only petroleum products, similar price increases occurred in coal over 1973-74. Energy prices move together—market competition tends to keep one fuel in a constant relative price relationship with the others. Even natural gas, which is partially regulated, has risen over 30 percent. Thus, contrary to the conventional wisdom among many hydrogen economy proponents, it is unlikely that coal will become a "cheap" source of hydrogen even if petroleum prices continue to increase, and it is unlikely that hydrogen will become less costly than conventional fuels.

The future costs of fuels from coal were addressed in the Project Independence study by the Synthetic Fuels Task Force [6]. The cost estimates for SNG were recalculated, using the 20 percent fixed charge factor used elsewhere in this study, and these estimates are presented in Table 11.11. With advanced technology, the costs are considerably above Linde's estimate for hydrogen from coal at $1.60 per $10^6$ Btu (see Table 11.4). From a systems standpoint, however, SNG has an advantage, since the transmission and distribution infrastructure for SNG already exists, and it can be incorporated into the existing natural gas system with no impact on commercial or residential users.

TABLE 11.11

Mine Mouth SNG Costs

|  | SNG Production Cost (per $10^6$ Btu) |
|---|---|
| Advanced technology | |
|   Eastern underground coal @ $16 per ton | $2.84 |
|   Western surface coal @ $5 per ton | 2.22 |
| Current technology (Lurgi) | |
|   Western surface coal @ $5 per ton | 2.60 |

Source: Adapted from reference 6.

The economics of alternative coal gasification schemes are such that doubling of coal prices does not affect their relative economic ranking as shown in Figure 11.5. A similar result occurs when capital costs change; the alternative processes have similar capital requirements in terms of dollar amounts and type of equipment, so that their relative ranking remains the same.

## GOVERNMENTAL ACTIONS IN ENERGY PRICING

Assessing the cost of hydrogen relative to alternative fuels is complicated by the role of governmental policies in the eventual outcome. Past energy prices have been greatly distorted by governmental regulation, as is demonstrated by the disparity between prices of regulated natural gas and unregulated intrastate natural gas. Given the current high level of concern with energy, it can be assumed that the federal government will continue to play a big part in future energy prices.

Currently, policy choices being considered are governed by the objectives of Project Independence. One can assume that the eventual policies will contain a combination of incentives to develop new energy sources and incentives to conserve energy. Stakeholders on all sides can be expected to argue their positions, seeking to prevent radical measures such as high excise taxes on gasoline or limiting imports to some past level. The long-run outcome concerning new energy sources will probably be determined by the government allocation of research and development funds among nuclear, solar, geothermal, and other technologies.

FIGURE 11.5

Synthetic Fuels Price Sensitivity to Coal Feed Costs*

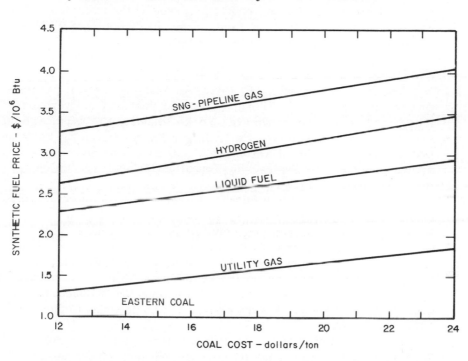

*Including a 15 percent discounted cash flow return on investment.

Source: Reference 6; hydrogen cost is based on reference 2.

REFERENCES

1. "Hydrogen and Other Synthetic Fuels," a summary of the work of the Synthetic Fuels Panel, prepared for the Federal Council on Science and Technology R&D Goals Study, September 1972.

2. J. E. Johnson, "The Economics of Liquid Hydrogen Supply for Air Transportation," in Advances in Cryogenic Engineering, vol. 19, ed. K. D. Timmerhaus (New York: Plenum, 1974), pp. 12-22.

3. E. Fein, "A Hydrogen Based Energy Economy," The Futures Group, Glastonbury, Conn., October 1972.

4. R. H. Wentorf, Jr., and R. E. Hanneman, "Thermochemical Hydrogen Generation," Science 185 (July 26, 1974): 311-19.

5. "The Crushing Burden of Corporate Debt," Business Week, October 12, 1974, p. 55.

6. "Synthetic Fuels from Coal," FEA Project Independence Blueprint Final Task Force Report, prepared by the Interagency Task Force on Synthetic Fuels from Coal, under the direction of the U.S. Department of the Interior, November 1974.

7. "Demonstrated Reserve Base," U.S. Bureau of Mines, U.S. Department of the Interior, 1974.

8. R. E. Chao and K. E. Cox, "An Analysis of Hydrogen Production Via Closed-Cycle Thermochemical Schemes," in Hydrogen Energy, ed. T. N. Veziroglu (New York: Plenum, 1975), pp. 317-30.

9. L. T. Blank et al., "A Hydrogen Energy Carrier," Systems Design Institute, National Aeronautics and Space Administration-American Society for Engineering Education, Johnson Space Center, Houston, Texas, 1973.

10. W. Hausz, "The Influence of Hydrogen in an Energy Marketplace," in Proceedings of the Cornell International Symposium and Workshop on the Hydrogen Economy, ed. S. Linke (Ithaca, N.Y.: Cornell University, April 1975), pp. 217-34.

11. D. P. Gregory et al., "A Hydrogen-Energy System," American Gas Association, Alexandria, Va., August 1972.

12. A. Konopka and J. Wurm, "Transmission of Gaseous Hydrogen," Ninth Intersociety Energy Conversion Engineering Conference, 1974, pp. 405-12.

13. H. D. Hottel and J. B. Howard, New Energy Technology, Some Facts and Assessments (Cambridge, Mass.: Massachusetts Institute of Technology Press, 1971).

14. J. E. Johnson, "Economics of Large Scale Liquid Hydrogen Production," paper presented at the Cryogenic Engineering Conference, Boulder, Colo., June 1966.

15. N. C. Hallet, "Study, Cost, and System Analysis of Liquid Hydrogen Production," National Aeronautics and Space Administration Circular no. 73226, June 1968.

16. J. E. Johnson, "The Storage and Transportation of Synthetic Fuels," a report to the Synthetic Fuels Panel working on the Federal Council on Science and Technology R&D Goals Study, September 1972.

17. Federal Power Commission News 7, no. 47 (November 22, 1974): 13-24.

18. J. F. Frawley, "1975-1985 Coal Industry's Capital Requirements," Mining Engineering, May 1975, pp. 38-40.

19. R. Shinnar et al., "Thermochemical Hydrogen Generation: Heat Requirements and Costs," Science 188 (June 6, 1975): 1036-38.

Hydrogen has diverse and wide-ranging applications both as an energy form and as a chemical. However, the investments required to produce, deliver, and use it are so large that deployment would have to be spread over many years. As a result, the societal consequences would also be felt over a period of many years.

This section first describes a tool useful in determining which applications of hydrogen are most sensible and could most readily be accommodated within the framework of the existing energy supply and use patterns of the United States. The tool is also useful in identifying key barriers that limit implementation. Hydrogen economy technologies are, therefore, characterized in a manner compatible with this tool. Second, transition scenarios depicting implementation under three generalized societal climates are presented.

# 12

## THE BUILDING-BLOCK CONCEPT

Technologies encounter physical limitations and exhibit economies of scale. The optimally sized unit, or "natural building block,"
characteristically has reaped nearly all of the economy-of-scale benefits possible and has been standardized in its manufacture. Rather
than build a device or system twice as large as the natural building
block, two building blocks are built in parallel. Good examples of
the natural building-block concept are railroad boxcars, petroleum
storage tanks, and 1,000-megawatt (equals 1 gigawatt, GW) nuclear
reactors.

The word "natural" associated with "building block" alludes to
physical, economic, managerial, or other relevant limitations on the
size of the building block that would be deployed in the real world.
The concept of natural building blocks can be illustrated by reference
to railroads: Freight cars do not vary much in size for several reasons. First, the weight loading that the rail can bear is an important
physical limitation. Second, there is a geometrical limitation on the
length of cars that can negotiate curves on existing track. Third, the
clearance in tunnels and along parallel tracks limits the height and
width of cars. Economies of scale associated with construction and
operation is an economic factor in the determination of the "natural"
size of freight cars, for once diminishing returns have set in heavily,
there is little incentive to increase freight car size. Similar considerations set the size and power of locomotives, which in turn determine the number of loaded freight cars a given locomotive can haul.
The combination of these natural building blocks dictates the form
and size of trains commonly seen. Thus, although a train consisting

of one locomotive and one freight car is conceivable, it is not practical (and seldom seen) because it violates the principle of achieving matches between natural building blocks.

Many hydrogen economy systems are conceivable, but they are not all equally practical. Analysis of hydrogen systems in terms of "natural" building blocks aids the determination of which among the conceivable systems is practical, and facilitates discovery of major mismatches between the building blocks composing a system, mismatches that generally would seriously impair deployment of a system.

## HYDROGEN ECONOMY BUILDING BLOCKS

The natural building blocks for hydrogen production, conversion, distribution, and demand that are most important to a hydrogen economy are given in Table 12.1. Because the output of one building block serves as the input to another block, the size of all building blocks has been expressed in a common unit that expresses the flow of energy in hydrogen form. The unit chosen is the gigawatt (equal to $10^9$, or 1 billion watts).

Application of the building-block analysis is illustrated in Figure 12.1, which depicts possible hydrogen systems for ammonia synthesis, fuel demand for subsonic passenger planes at a large airport, and residential gas use in a large city. Several important deductions can be made from these diagrams of conceivable hydrogen systems by reference to the natural building-block sizes shown in parentheses.

First, at the production end of the pipeline, a single nuclear electric/electrolytic hydrogen facility would be incapable of filling even the smallest reasonably sized trunk pipeline. Consequently, to operate the pipeline economically, a hydrogen input from about three nuclear power plants would be required. By contrast, one coal gasification building block matches well the capacity of a 24-inch diameter pipeline. However, it must be recognized that for the same length pipeline, operating at the same pressure of 1,000 psi, the cost of delivering one unit of hydrogen by 24-inch pipeline is about 50 percent higher than by 48-inch pipeline (with a capacity of 8.6 GW). Thus, although a building-block match can be made with coal gasification and a 24-inch pipeline, it is not the most economically favorable one.

Second, at the demand end of the pipeline, the ammonia synthesis building block would require only about 15 percent of the pipeline capacity and about 50 percent of the electrolytic hydrogen derived from a single nuclear electric power plant. By contrast, the airport would require more hydrogen than a 24-inch pipeline could deliver; indeed, a 30-inch pipeline would be required. In addition, the airport would require about five times the output of a single nuclear electrolysis

TABLE 12.1

Natural Building Blocks of a Hydrogen Economy

| Building Block | Size (energy flow in $H_2$ form, GW) |
|---|---|
| Production | |
| Nuclear fission electric generation | |
| Nominal electric output—1.0 (electricity) | |
| Output at 75 percent net availability factor— 0.75 (electricity) | |
| Electrolytic hydrogen at 70 percent efficiency | 0.53 |
| Nuclear thermochemical water splitting (rough estimate) | 1.5 |
| Coal gasification to hydrogen (1,000 tons of $H_2$ per day) | 1.5 |
| Liquefaction facility | |
| Largest experience (60 tons of $LH_2$ per day) | 0.076 |
| Projected (250 tons of $LH_2$ per day) | 0.32 |
| Distribution | |
| Gas in pipelines optimized for hydrogen (1,000 psi, $H_2$ used for fuel in pumping stations at a cost of $4 per $10^6$ Btu) | |
| Pipeline diameter | |
| 24 inch | 1.7 |
| 30 inch | 2.7 |
| 36 inch | 4.3 |
| 42 inch | 6.5 |
| 48 inch | 8.6 |
| Liquid in railroad cars (100 cars per train, 28,000 gallons per car, one train per day) | 1.0 |
| Demand | |
| Automobile using liquid hydrogen (per 100,000 population, 1971 fuel use rates, 2.2 people per car owned) | 0.13 |
| Residential gas (per 100,000 population, typical methane energy demand, 3.2 people per household) | 0.16 |
| Major airport (San Francisco, SFO, 1973) | 2.5 |
| Iron ore reduction (500,000 tons iron per year) | 0.20 |
| Ammonia synthesis (1,000 tons ammonia per day) | 0.26 |

FIGURE 12.1

Conceivable Hydrogen Economy Systems

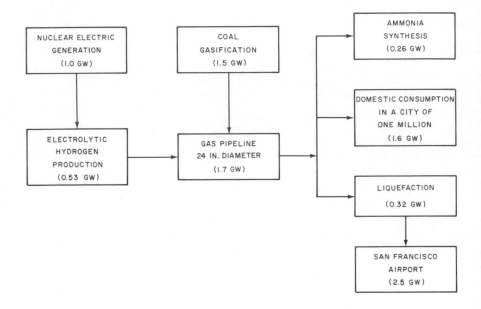

Note: Natural building block sizes in parentheses.


plant, and about eight times the capacity of a single large liquefaction
plant.  It is questionable whether five nuclear power plants could be
sited closely enough to feed a 30-inch pipeline efficiently.  When multiples of the building-block units are needed, however, the system reliability increases because it is then less likely that an accident or
malfunction could cripple the entire system.  Moreover, the use of
multiple building blocks adds flexibility for scheduled maintenance
and inspection shutdowns.

   Third, a city of 1 million inhabitants that used hydrogen domestically would require enough hydrogen to match, and thereby to justify,
a single 24-inch pipeline and a single coal gasification plant, while
about three nuclear electrolytic plants would be needed to supply the
same demand.  There are about 35 Standard Metropolitan Statistical
Areas (SMSAs) in the United States with population in excess of 1 million.

   Although the building-block sizes are called "natural," they are
not entirely immutable, and an actual facility could be built to handle

somewhat different flows. Unfortunately, however, many of the mismatches shown in Figure 12.1 and discussed above exceed the flexibility of a building block. Components of the hydrogen economy cannot be expected to be deployed in uneconomical sizes that fail to yield a return on investment comparable to alternative uses of capital. Hydrogen applications with small-demand building blocks must usually wait until other small-demand uses are ready to be deployed, for then, together, they can justify deployment of the larger supply building blocks that otherwise are inhibited. On the other hand, provided that reasonable multiples of building blocks are required, some use sectors can justify deployment earlier if their demand exceeds the "natural" capacity of production and distribution building blocks. These end-uses alone can justify the larger system. Occasionally, slight spare capacity in these large systems could result in a sharing with a small nearby end-use, thereby somewhat accelerating deployment of relatively small-scale hydrogen uses.

As demonstrated by the discussion above, building-block analysis can identify the systems that are physically and economically the most likely to be deployed. The approach can also identify synergisms that influence the likely order of deployment. Additional analysis is required, however, to determine whether these systems are feasible in terms of the existing institutional framework of corporations and governments.

# CHAPTER 13

## TRANSITION SCENARIOS

### IMPORTANT NONECONOMIC FACTORS

Scenarios of a transition to a hydrogen economy must be based upon six important considerations:

1. Status of the relevant technology (conceptual, experimental, developmental, proven, and so on)
2. Lead time necessary to deploy the technology once it is proven
3. Temper of the decision-making climate and the time required before it might become favorably disposed toward hydrogen
4. Natural building-block matches and mismatches
5. Synergisms among end-use sectors
6. Allocation priorities for dwindling supplies of natural gas

The last consideration is an especially important determinant of the potential evolution of the market for hydrogen used as a gaseous fuel. Recent governmental allocation priorities suggest that, as supplies of natural gas (plus methane SNG) dwindle, consumers will be denied consumption in roughly the following order:*

1. Industry that burns methane for heat, including electric generation
2. Industry that uses methane as a chemical source of hydrogen (except ammonia producers)

---

*See Appendix to Chapter 16 for a more complete listing of the current Federal Power Commission curtailment priorities.

   3.  Petrochemical industries
   4.  Ammonia producers
   5.  Residential and commercial consumers

Residential and commercial use will be protected the longest because
the number of people directly affected is the greatest and there are
few alternatives that would be easy to implement.  Ammonia producers
are likely to be the next most favored because their product is crucial
to agriculture, and a major dislocation in ammonia supplies could be
disastrous to the food supply, food prices, and the U.S. balance of
trade.  Even though many jobs would be affected if industries were
forced to shut down for lack of fuel, industries generally have the
most flexibility in converting to alternative fuels.
   A seventh important consideration, the relative cost of energy
in the form of hydrogen, has not been factored into these scenarios
explicitly for three reasons:

   1.  Hydrogen must necessarily cost more than the primary or
secondary energy forms from which it is derived.
   2.  Interfuel competition, which is most intense in the kind of
tight energy supply situation now facing the United States, tends to
balance the unit cost of the various forms of energy; thus it is extremely unlikely that the cost of hydrogen could undercut the price of
its major competitors.
   3.  Ignoring the relative cost factor (which could only delay interest in hydrogen), the time when hydrogen implementation could
reach a significant scale is so distant that any attempt to predict energy prices and relative market shares for that time would be mean-
ingless.


                              SCENARIOS

   The scenarios presented in Figures 13.1 to 13.3 depict three
generalized decision-making climates:

   1.  Optimistic, assuming no government intervention, ignoring
hydrogen's economic disadvantage, and assuming minimum decision
and technology lead times
   2.  Realistic, assuming some government involvement in decision making, some improvement in hydrogen's cost relative to alternatives, and less optimistic lead times
   3.  Strong government intervention, federal government decree
of the use of hydrogen, backed by appropriate barriers and incentives.

# FIGURE 13.1

## Optimistic Implementation Scenario

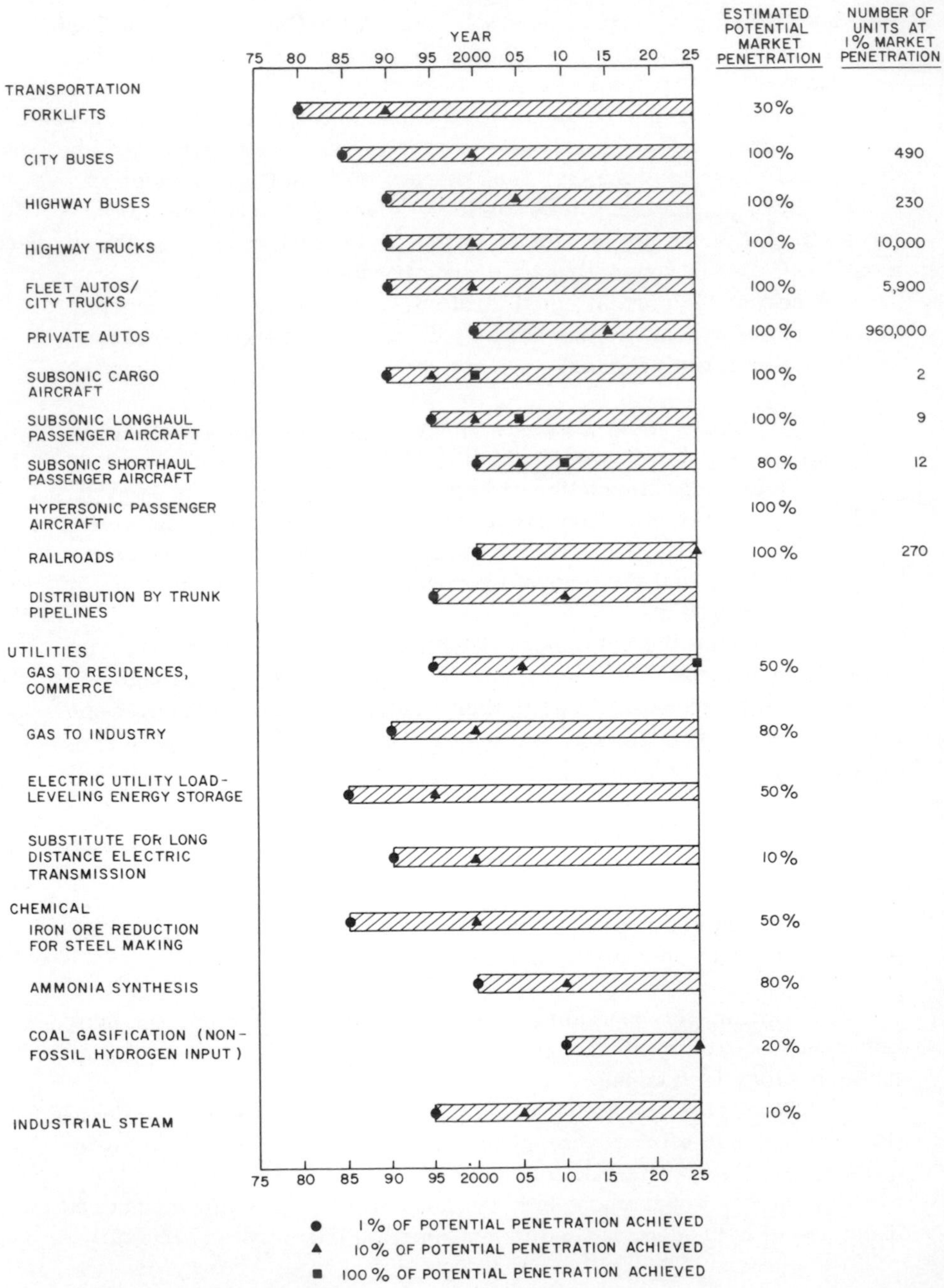

| | YEAR (75 80 85 90 95 2000 05 10 15 20 25) | ESTIMATED POTENTIAL MARKET PENETRATION | NUMBER OF UNITS AT 1% MARKET PENETRATION |
|---|---|---|---|
| **TRANSPORTATION** | | | |
| FORKLIFTS | | 30% | |
| CITY BUSES | | 100% | 490 |
| HIGHWAY BUSES | | 100% | 230 |
| HIGHWAY TRUCKS | | 100% | 10,000 |
| FLEET AUTOS/ CITY TRUCKS | | 100% | 5,900 |
| PRIVATE AUTOS | | 100% | 960,000 |
| SUBSONIC CARGO AIRCRAFT | | 100% | 2 |
| SUBSONIC LONGHAUL PASSENGER AIRCRAFT | | 100% | 9 |
| SUBSONIC SHORTHAUL PASSENGER AIRCRAFT | | 80% | 12 |
| HYPERSONIC PASSENGER AIRCRAFT | | 100% | |
| RAILROADS | | 100% | 270 |
| DISTRIBUTION BY TRUNK PIPELINES | | | |
| **UTILITIES** | | | |
| GAS TO RESIDENCES, COMMERCE | | 50% | |
| GAS TO INDUSTRY | | 80% | |
| ELECTRIC UTILITY LOAD-LEVELING ENERGY STORAGE | | 50% | |
| SUBSTITUTE FOR LONG DISTANCE ELECTRIC TRANSMISSION | | 10% | |
| **CHEMICAL** | | | |
| IRON ORE REDUCTION FOR STEEL MAKING | | 50% | |
| AMMONIA SYNTHESIS | | 80% | |
| COAL GASIFICATION (NON-FOSSIL HYDROGEN INPUT) | | 20% | |
| INDUSTRIAL STEAM | | 10% | |

● 1% OF POTENTIAL PENETRATION ACHIEVED

▲ 10% OF POTENTIAL PENETRATION ACHIEVED

■ 100% OF POTENTIAL PENETRATION ACHIEVED

172

# FIGURE 13.2

## Realistic Implementation Scenario

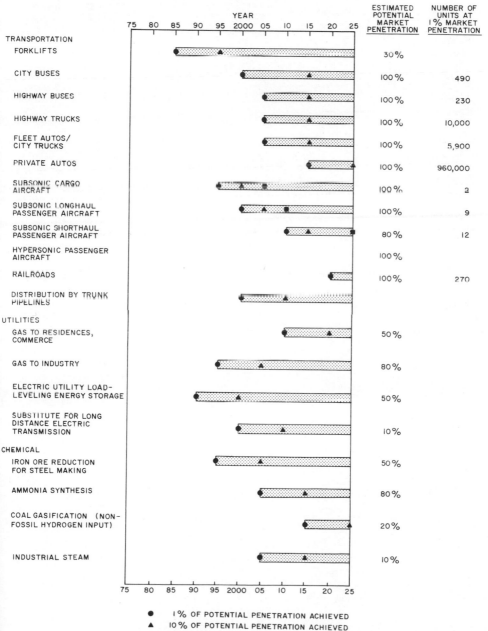

| | | ESTIMATED POTENTIAL MARKET PENETRATION | NUMBER OF UNITS AT 1% MARKET PENETRATION |
|---|---|---|---|
| **TRANSPORTATION** | | | |
| FORKLIFTS | | 30% | |
| CITY BUSES | | 100% | 490 |
| HIGHWAY BUSES | | 100% | 230 |
| HIGHWAY TRUCKS | | 100% | 10,000 |
| FLEET AUTOS/ CITY TRUCKS | | 100% | 5,900 |
| PRIVATE AUTOS | | 100% | 960,000 |
| SUBSONIC CARGO AIRCRAFT | | 100% | 2 |
| SUBSONIC LONGHAUL PASSENGER AIRCRAFT | | 100% | 9 |
| SUBSONIC SHORTHAUL PASSENGER AIRCRAFT | | 80% | 12 |
| HYPERSONIC PASSENGER AIRCRAFT | | 100% | |
| RAILROADS | | 100% | 270 |
| DISTRIBUTION BY TRUNK PIPELINES | | | |
| **UTILITIES** | | | |
| GAS TO RESIDENCES, COMMERCE | | 50% | |
| GAS TO INDUSTRY | | 80% | |
| ELECTRIC UTILITY LOAD- LEVELING ENERGY STORAGE | | 50% | |
| SUBSTITUTE FOR LONG DISTANCE ELECTRIC TRANSMISSION | | 10% | |
| **CHEMICAL** | | | |
| IRON ORE REDUCTION FOR STEEL MAKING | | 50% | |
| AMMONIA SYNTHESIS | | 80% | |
| COAL GASIFICATION (NON- FOSSIL HYDROGEN INPUT) | | 20% | |
| INDUSTRIAL STEAM | | 10% | |

● 1% OF POTENTIAL PENETRATION ACHIEVED
▲ 10% OF POTENTIAL PENETRATION ACHIEVED
■ 100% OF POTENTIAL PENETRATION ACHIEVED

FIGURE 13.3

Strong Government Intervention Implementation Scenario

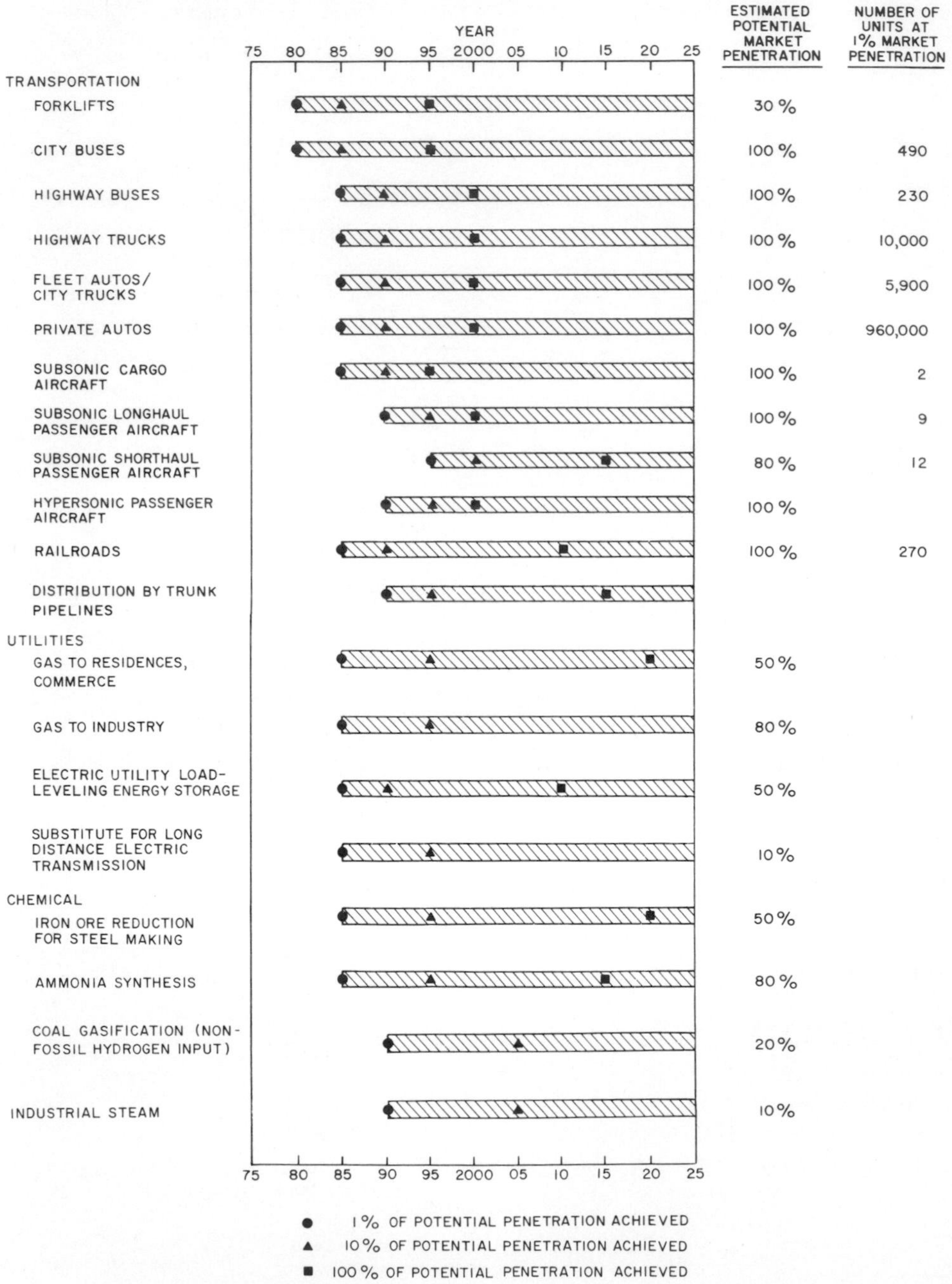
YEAR
75 80 85 90 95 2000 05 10 15 20 25
ESTIMATED POTENTIAL MARKET PENETRATION
NUMBER OF UNITS AT 1% MARKET PENETRATION
TRANSPORTATION
FORKLIFTS 30%
CITY BUSES 100% 490
HIGHWAY BUSES 100% 230
HIGHWAY TRUCKS 100% 10,000
FLEET AUTOS/ CITY TRUCKS 100% 5,900
PRIVATE AUTOS 100% 960,000
SUBSONIC CARGO AIRCRAFT 100% 2
SUBSONIC LONGHAUL PASSENGER AIRCRAFT 100% 9
SUBSONIC SHORTHAUL PASSENGER AIRCRAFT 80% 12
HYPERSONIC PASSENGER AIRCRAFT 100%
RAILROADS 100% 270
DISTRIBUTION BY TRUNK PIPELINES
UTILITIES
GAS TO RESIDENCES, COMMERCE 50%
GAS TO INDUSTRY 80%
ELECTRIC UTILITY LOAD-LEVELING ENERGY STORAGE 50%
SUBSTITUTE FOR LONG DISTANCE ELECTRIC TRANSMISSION 10%
CHEMICAL
IRON ORE REDUCTION FOR STEEL MAKING 50%
AMMONIA SYNTHESIS 80%
COAL GASIFICATION (NON-FOSSIL HYDROGEN INPUT) 20%
INDUSTRIAL STEAM 10%
75 80 85 90 95 2000 05 10 15 20 25
● 1% OF POTENTIAL PENETRATION ACHIEVED
▲ 10% OF POTENTIAL PENETRATION ACHIEVED
■ 100% OF POTENTIAL PENETRATION ACHIEVED

The scenarios do not necessarily indicate what will happen but only what could happen under those situations. For example, in the automotive sector, the development of synthetic gasoline and diesel fuel derived from coal and oil shale is likely to preempt (for many decades) the role hydrogen could conceivably play.

In Figure 13.4 the market penetration under the three scenarios is shown on one diagram for selected end-uses. This presentation permits a rapid evaluation of the importance of policy to the different end-uses. The aircraft sector exhibits a complete transition over 10 years under all policy climates. By contrast, a strong government policy is seen as necessary if widespread use of hydrogen in private autos is to occur in the next 50 years. Thus, in those sectors where there are many small users and the attraction (for example, environmental) of hydrogen is outside existing market mechanisms, strong government intervention would be required to force adoption of hydrogen.

Figure 13.5 shows the time between 1975 and 2025 divided into five eras of the hydrogen economy evolution:

1. Incremental additions to existing production and distribution practices
2. Stand-alone production and consumption
3. Sparse, nearly single-purpose distribution networks
4. Embryonic forms of generalized distribution networks
5. Mature distribution network that becomes continually more dense

End-uses likely to emerge within each era are shown; an end-use is shown when market penetration reaches 1 percent of its ultimate potential, and thus its appearance does not represent full-scale use. Because some uses become more likely once other uses are under way, a key factor is the occurrence of synergistic effects among end-uses.

The logic behind the three scenarios and the sequence of hydrogen evolution can be summarized as follows:*

1. Very small-scale, special-purpose, special-need uses could begin at once if existing production and distribution practices were used.

2. Some fairly large demonstration-type experiments could also proceed if the existing production and distribution systems were augmented.

---

*More elaborate discussion of pacing factors and obstacles can be found in Chapter 14 (automotive), Chapter 15 (aviation), Chapter 16 (utilities), and Chapter 17 (steelmaking and ammonia synthesis).

# FIGURE 13.4

## Comparison of Scenarios for Selected End-Use Sectors

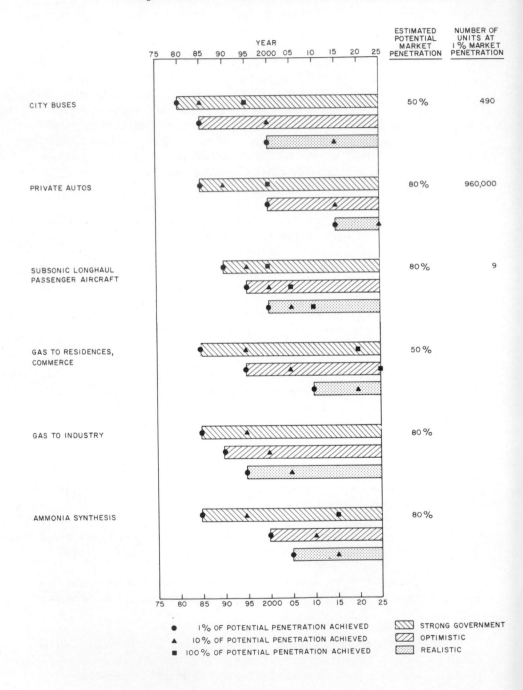

# FIGURE 13.5

## Eras and Events in the Transition to a Hydrogen Economy

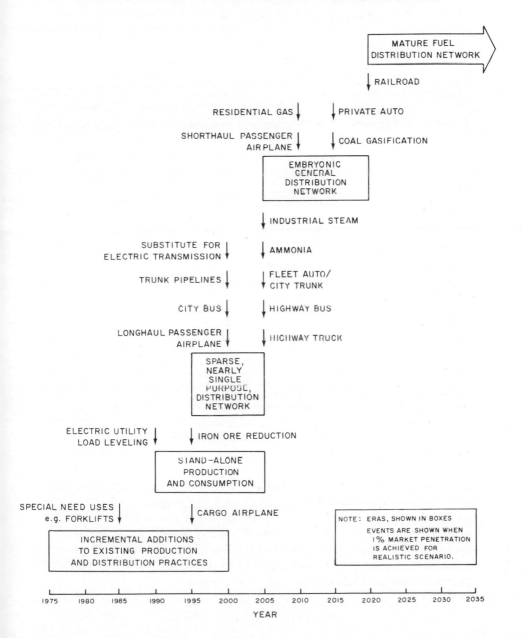

3.  Large-scale, stand-alone hydrogen production for internal
(captive) consumption could proceed whenever the state of the art of
the relevant technologies becomes more advanced; such use, however,
does not increment the production and distribution system of "merchant hydrogen," but it does advance the state of the art of production,
distribution, and storage.

4.  An advanced state of the art and increased interest in hydrogen could lead to deployment of limited-capacity distribution systems
in narrowly defined geographic areas to serve large, steady customers.

5.  Small-scale distribution networks could serve as the basis
for the enlargement of production and distribution networks; advanc-
ing hydrogen economy state of the art could make applications that
depend on dense distribution networks increasingly attractive; the process would then begin to gain momentum with much cross-sector
stimulation.

6.  A mature, widespread distribution network would begin to
grow; and hydrogen production and consumption could proliferate.

Whether a state of affairs resembling the scenarios presented here ever
actually evolves depends on myriad future decisions made by a vast
number of individual consumers, corporate decision makers, investors, government policy makers, and government administrators and
regulators.  History suggests that each incremental decision will be
based mainly on the economics and politics of the moment and of a few
subsequent moments rather than upon lofty philosophical issues.  Thus,
at each decision-making juncture, it will be more important whether
hydrogen is the economically correct decision than whether hydrogen
is the "good thing" that enthusiasts assert.  The effects of these decisions will accumulate to determine the future role of hydrogen.

The horizons for individual, corporate, and governmental planning are not distant.  They rarely extend beyond five years in any
meaningful way.  All decision makers intuitively and informally discount the future in much the same way as economists discount future
cash flows with an explicit discount rate.  For example, if what happens next year is consistently judged to be 75 percent as important as
what happens this year, then what happens ten years from now is only
valued at 6 percent (0.75 raised to the tenth power).  Maxims such
as "Live for today" and "A bird in the hand is worth two in the bush"
suggest that few individuals accord to even the next year an informal
value as high as 75 percent.

## SUMMARY AND DISCUSSION

If the hydrogen economy is ever to take form, passage through
a series of steps at least qualitatively similar to those shown in Fig-

ure 13.5 seems inescapable.  The process must necessarily be protracted because there are long lead times for each successive step,
and huge investments of capital and labor.  Consequently, even
within the relatively distant future, the hydrogen economy could not
come to full flower and, for a very long time, the relevant societal
impacts are those that result from the transition process as opposed
to those that would result from reaching a mature hydrogen economy.

Although it can be argued convincingly that a combination elec-
tric/hydrogen economy is ultimately inevitable, there is no assurance
that the weight of the many incremental individual corporate and governmental decisions will lead society in that direction within the next
century.  Indeed, for the rest of this century there are so many energy
carrier alternatives that blend more readily into the existing order of
things that there is a strong possibility that society will "slog" through
all the fossil-fuel alternatives before settling on hydrogen.

In summary:

1.  The total lead time for a hydrogen economy is very long.

2.  Most of the first major uses will involve "captive," rather
than "merchant" hydrogen.

3.  In the year 2000 the importance of the hydrogen economy
will be measured mainly by whether the United States is on a course
toward it or away from it.

4.  Total hydrogen use, even by the year 2025, would be small
compared to the full potential demand.

5.  Yet, in a few sectors, notably commercial passenger aviation, hydrogen use could reach maturity by 2025.

Hydrogen could be used in essentially every application requiring energy and in some chemical applications as well. The societal implications of such uses, however, vary widely in their gravity. This section focuses on four areas judged to have the largest, most widely distributed potential consequences: automotive, aviation, utilities, and steelmaking and ammonia synthesis.

# 14

## IMPACTS OF HYDROGEN-FUELED
## PRIVATE AND FLEET
## AUTOMOTIVE VEHICLES

PRIVATE AUTOMOBILES

The System and the Stakeholders

Privately owned automobiles are but one part of a complex system of automotive transportation. The complete system consists of vehicles; fuel producers, distributors, and retailers; vehicle manufactures and sales outlets; vehicle maintenance and repair facilities; and roadways. The last four items can be viewed as supporting or infrastructure networks, and, as far as hydrogen is concerned, the most critical is the fuel logistics network. Except for roadways, this is the network with which the vehicle owner must most often interact. As a consequence, the outlets for this network have become very much more widespread than the elements of the other infrastructure networks.*

It is characteristic of many networks that once they are widely deployed, their standards become fixed and they become resistant to change and, therefore, extremely difficult to supplant. To a very real extent, networks become the dominant aspect of systems, and the concerted effort and expense needed to make even minor changes deflect many conceivable improvements. There are many examples of institutionalized networks that seemingly will impose their standards on U.S. society indefinitely, such as the gauge (separation) of railroad tracks, the 60-Hz, 115-volt electrical system, the transmission stan-

---

*Although many fuel filling stations also offer some maintenance and repair service.

dards of broadcast television, and the bandwidth of telephone circuits. For better or for worse, these networks and their standards evolved and became established when there was no competitor or even near competitor to the system they represent. Yet, once established and widely used, they exert a practical limit on future possibilities.

A system of automobiles equipped with either liquid hydrogen or metal hydride fuel storage is simple to imagine, but the supporting fuel distribution network implied is quite different from the present gasoline logistics network and by no means represents a small, evolutionary change of that network. Rather, hydrogen would require a completely different system, and a transition from our present gasoline system to a hydrogen system would, therefore, require an enormous degree of effort, expense, and coordination. Indeed, so large are these inhibiting factors that the fuel logistics network is probably the single most important barrier to the realization of a hydrogen-fueled automotive system. This assertion deserves further examination because the principles involved are also applicable to the potential changeover to hydrogen in other energy-using systems considered elsewhere in this report.

To the consumer, a vehicle is, first of all, a source of personal mobility. Any change in the automotive system that would detract from his accustomed or expected freedom of mobility would diminish the attractiveness of owning a car. If, for example, fuel stations were to become less numerous and farther apart, or if the availability of his particular fuel were reduced, then the owner's convenience would also be reduced and along with it his satisfaction with car ownership. Two examples illustrate this point.

First, diesel-powered cars have long been available (although at a higher price than the equivalent gasoline-powered car), but sales have never been large [1]. The low sales probably stem largely from the consumer's reluctance to commit himself to dependence on the diesel fuel network,* which is substantially more sparse (mainly truck stops or marinas) than the nearly ubiquitous gasoline network. Consumers have shown preference for gasoline-powered cars, even though diesel cars both get better mileage and usually have larger fuel tanks (thereby increasing the range between refueling).

---

*Some will argue that the sluggish performance of diesel-powered cars compared to gasoline-powered cars is the "real" reason diesel cars have not been popular. We acknowledge this contribution but doubt that this is the main reason; many gasoline-powered vehicles, such as the Volkswagen buses, also exhibit poor performance but nevertheless have sold well.

Second, the recent introduction of standards to tighten automobile air pollutant emissions has led to the use of catalytic converters to clean carbon monoxide and hydrocarbons from car exhaust. Unfortunately, the lead-containing chemical compounds routinely added to most gasolines to improve their octane rating (and thus the smoothness of combustion) rather quickly "poison" the platinum catalysts in these converters, rendering them useless after as little as two tanks of leaded fuel. Consequently, the EPA mandated that lead-free fuel be made available nationwide to protect the converters from degradation [2]. To assure adequate availability of lead-free fuel involved considerable negotiations over a period of several years between EPA and the fuel suppliers concerning the rules defining how many and which gasoline filling stations would be required to carry the lead-free fuel [2].

Meanwhile, individual consumers and automotive interest groups (such as the American Automobile Association) expressed concern about the availability of lead-free fuel in rural areas and in Canada and Mexico, while automobile manufacturers worried that these consumer concerns about fuel availability would lead to low sales of the new catalyst-equipped cars. Without question, the complexity of the change implied by a transition to hydrogen far exceeds the recent partial change to lead-free gasoline; and although the lead-free gasoline issue has now been resolved successfully, the degree of controversy engendered indicates what may be expected should a change to hydrogen begin to take form.

Because of the kinds of problems indicated above, it generally proves far simpler to change the nature of the vehicle that interacts with an unchanged fuel system than to change the fuel system itself. A good example of this is the way Mazda was able to introduce the Wankel, or rotary, engine. Because the buyer's mobility and refueling habits were not threatened, little resistance to engine innovation arose.* Far more radical engine changes could be readily accomplished as long as the engines used the same fuel and thus did not threaten to impair the consumers' mobility.

Consumers, of course, are not the only parties with a stake in a transition in the fuel network. Fuel vendors have a financial stake that far exceeds that of the individual consumer. There are basically three classes of gasoline station operations:

------

*The consumer did have to be concerned about the adequacy of the repair and maintenance network, but since he usually interacts with this network far less frequently than with the fuel network, his concern was less acute.

1.  Stations owned by the oil company whose product is sold

2.  Independently owned stations franchised by the oil company whose brand is sold

3.  Independent stations that purchase wholesale from several producers

Since the first two classes are by far the dominant business forms dispensing gasoline to the public, it is basically the decisions of the major oil corporations rather than the station operator that determine the fuel to be sold.

As long as the major oil companies perceive that they can continue to supply gasoline in basically the quantities desired by consumers, they naturally have very little interest in embarking on a transition to hydrogen.  Even when they begin to question their capability of sustaining supplies from conventional (natural) crude oil from either domestic or foreign sources, they can be expected first to turn their attention to synthetic crude oils derived from oil shale and coal [3-5].  These synthetics hold several attractions to the existing major oil companies not offered by hydrogen.  First, synthetic crudes can be blended with conventional  crudes, thereby protecting existing investments in facilities for transportation, refining, and marketing from premature obsolescence [3, 5].  Second, companies, like the individuals that compose them, tend to prefer to do those things that are familiar.  As a result, slowly transforming the business from dependence on conventional sources of oil to synthetic oils is a far more comfortable prospect than blazing a trail to hydrogen [5].  Moreover, with an uncertain market for hydrogen but a certain market for gasoline, the major oil companies have shown (and will likely continue to show) very little interest in hydrogen-fueled automobiles.

It is conceivable, of course, that an aggressive company presently outside the automotive fuels market might decide to offer hydrogen through a special chain of fuel outlets in competition with gasoline.  Probably the most appropriate corporate interests here are large chemical companies (as evidenced by Union Carbide's interest in hydrogen) because of their large resources and technical know-how.  The success of this conceivable course is unlikely, however, because it would involve vigorous competition between products and companies rather than merely a gentle competition between products (such as gasoline and diesel) of the same company.  The high cost of marketing would be a great disadvantage to the newcomer because of the uneven match of assets and the automotive marketing wherewithal of the rival companies [5].

Automobile makers also would have a large stake in any future hydrogen automobile system and would naturally be concerned whether consumers would buy a hydrogen-fueled automobile and whether fuel would be available. All the concerns in the no-lead gasoline experience would apply, but in heightened form. In addition, automobile makers would be concerned that the small initial production runs of hydrogen-powered cars would make it difficult to reap economies of scale and thus would either mean low profits on hydrogen cars or else mean sales at noncompetitive prices.

## The Dilemma

Because neither the fuel producer, the auto maker, nor the consumer is especially eager to take the first risky step, a transition to a hydrogen-fueled automotive system poses a kind of three-way dilemma. This greatly stifles a free-market evolution from a gasoline-based to a hydrogen-based automotive system. It is natural to ask, then, how the present automotive system evolved and how its strong, and presently confining, inertia became established in the first place.

Historically, the automotive system—both vehicles and fuel—evolved together from a base of zero [1]. Indeed, the strong market interdependence between the auto makers and the oil companies is traceable to their simultaneous and symbiotic growth. When both started, there were many highly competitive auto-making and oil companies, but business failures and consolidations led both the vehicle and fuel production aspects of the automotive system to become far more concentrated and oligopolistic. The horse-powered transportation mode that gasoline-powered automobiles displaced was likewise fragmented, with no meaningful consolidation of market power in wagon makers, horse breeders, or hay sellers. Moreover, because the automobile competitively offered an increase in sustainable speed that the horse could not offer, the transportation modes were differentiated in quality and not really equivalent.

Moreover, there is an important asymmetry in the three-way dilemma: On the vehicle-manufacturing and fuel-marketing sides of the triangle the decision making is relatively concentrated in a few individuals, but on the consumer side the decision making is highly dispersed. In particular, while the decision to make a hydrogen-fueled vehicle resides in the president or board of directors of major automobile makers and the decision to produce the fuel similarly lies in the upper echelons of energy company executives, in a free market the decision whether and when to buy a hydrogen-fueled car would reside with literally millions of individual consumers. For exactly this reason, without governmental forcing of the issue, the prospect for hydrogen-fueled fleet vehicles is far greater than for personal vehicles. Accordingly, the transition scenarios of Chapter 13 showed the advent of hydrogen-powered fleet vehicles considerably earlier than that of personal vehicles.

## Transition Strategies

The considerations discussed above are general and would be most acutely felt if a transition to hydrogen-fueled vehicles were to occur simultaneously throughout the nation. Coordinated decision making between vehicle makers and fuel suppliers would be essential and would almost certainly have to follow the use of hydrogen in other sectors, both to facilitate fuel distribution and to provide valid experience on which consumer reaction to hydrogen could be based.

In a nationwide program, the fuel suppliers' strategy would probably be to provide hydrogen at only a few centrally placed locations within metropolitan regions and await an increased number of vehicles before opening additional fueling stations. Consumer inconvenience in dealing with only a limited number of fueling points would surely impede vehicle sales for an indefinite period, but if momentum could be established, then the transition would tend to be self-reinforcing: more vehicles would justify more fuel stations, which would lead to more vehicles, and so on. Indeed, at some point there would be a crossover at which the diminished number of gasoline-powered cars would, in turn, have so reduced the number of gasoline stations that hydrogen rather than gasoline would become the more attractive fuel to consumers. There is the implication, however, that midway through the transition, neither fuel network would be as extensive as people now experience, and thus there would be a temporary reduction in the net convenience of both forms of personal automotive travel.* The major problem of course, is how the transition process could be begun and sustained (if the mechanism of transition were easy, the diesel-powered car might already have taken over).

It is sometimes suggested that hydrogen-powered vehicles might be introduced by mandate in a relatively small region that has severe air pollution as a means to improve air quality. Setting aside the complex political question of whether such a mandate could ever become feasible in a democratic U.S. society, this approach would greatly offset the inherent difficulties of transition. In particular, because the number of hydrogen-powered vehicles in the region would grow quickly, the deployment of fueling stations would proceed quickly, thereby alleviating consumer inconvenience. While this would improve flexibility of travel within the region, the usefulness of the hydrogen-powered car might very well be limited to the mandated region. Perhaps expanded and subsidized use of gasoline-powered rental cars would provide an acceptable mechanism for travel outside the mandated region whenever necessary. Travel into the restricted region by mi-

---

*During this era people might shift increasingly to public transit.

grants and tourists and a residual population of aged gasoline-powered cars would require either the indefinite maintenance of a skeletal gasoline distribution network or the use of rental hydrogen-powered cars by visitors.

## Air Quality Implications

A complete transition to hydrogen-powered automobiles would have an enormous impact on urban air quality because the automobile is the dominant source of air pollutants in most cities. However, as suggested in the transition scenarios of Chapter 13, the time for automotive transition is distant and the transition would proceed slowly. Accordingly, improvements in air quality (if any) could only be realized slowly. Even if the transition were to be mandated tomorrow and were to proceed as quickly as new cars could be built, the period of transition would last for more than ten years because the annual turnover in vehicles runs about ten percent per year [7].

The clean air benefits of hydrogen-powered cars have received considerable attention from researchers [8-13], hobbyists [14, 15], and the press [16-20]. Figure 14.1 shows a widely quoted test result (obtained on a small specially adapted engine) that reports nitrogen oxide emissions considerably below those for gasoline [13]. * Such low levels of nitrogen oxide emissions are not universally reported, however. For example, Figure 14.2 reports, to the contrary, that at power output levels suitable for automative propulsion, the nitrogen oxides from hydrogen and gasoline are similar [21]. Because nitrogen oxide formation is mainly related to the peak temperatures achieved during combustion, there is apparently no intrinsic advantage in a hydrogen-air engine with regard to these emissions.† There is presently debate among hydrogen engine researchers concerning conflicting nitrogen oxide data exemplified by Figures 14.1 and 14.2 [22].

Although hydrogen-powered cars might prove to be nearly pollution-free, the building-block analysis of Chapter 12 shows that considerable industrial activity would occur where the hydrogen was generated. Therefore, because of this behind-the-scenes industrial activity, the locus of pollutant emissions would be transferred from streets and

---

*Since hydrogen contains neither carbon nor hydrocarbons, these pollutants would not be expected. This is the theoretical basis for the attraction of hydrogen as a motor fuel. However, trace levels to substantial amounts of these emissions can derive from the consumption of lubricating oil.

†A hydrogen-oxygen engine could not produce nitrogen oxides.

FIGURE 14.1

Nitrogen Oxide (NO$_x$) Emission from a Small Engine as a Function of Fuel Type

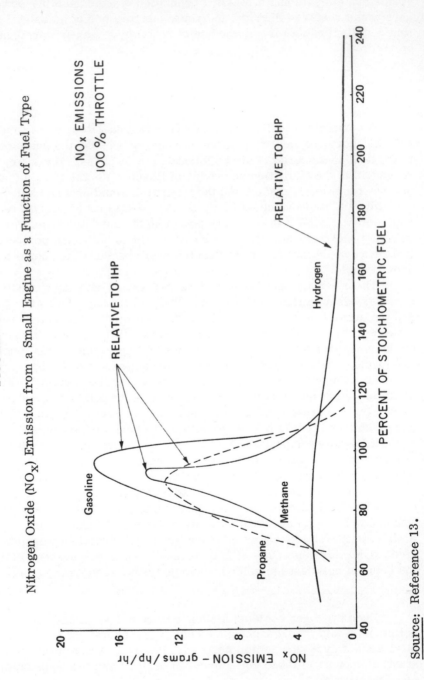

Source: Reference 13.

190

FIGURE 14.2

Nitrogen Oxide ($NO_x$) Emission from a Hydrogen-Fueled Automobile
Engine Compared to Same Engine Fueled with Gasoline

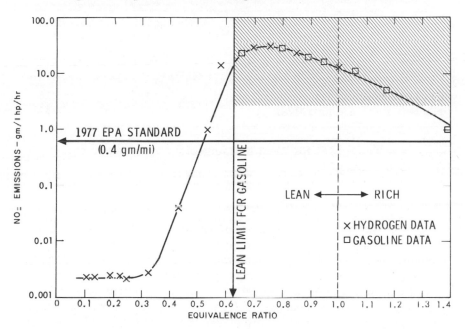

Note: Hatched area corresponds to power levels suitable for
automotive propulsion.

Source: Reference 21.

roads to hydrogen-generating plants elsewhere. Although these would
probably emit less "total" pollution (by some measure) than corres-
ponding travel in gasoline-powered automobiles, the pollutants may
be of an entirely different species. For example, if hydrogen were
generated either electrolytically or thermochemically from a primary
energy source, the net effect would be to exchange carbon monoxide
and hydrocarbon emissions for nuclear power thermal discharges into
bodies of water or air used to cool the reactor, low-level radioactive
emissions [23], the generation of radioactive waste, and the threat of
theft of dangerous materials [24] (such as plutonium). These problems of using nuclear energy for hydrogen production provide strong
arguments in favor of using solar energy technologies instead.

TABLE 14.1

Hydrogen Generation Needed to Supply 1973 California Automobile
Demand

| | Number of Building-Block-Sized Plants[a] at Motor Thermal Efficiency | |
| --- | --- | --- |
| | Equal to Gasoline | Enhanced[b] |
| Coal gasification | 21 | 16 |
| Nuclear fission/electrolysis | 58 | 44 |

[a]See Table 12.1 for plant building-block size and output.
[b]See Figure 7.7.

To illustrate, automobile registration in California was about
11 million cars in 1973. To produce enough hydrogen to power these
vehicles would require hydrogen generation as shown in Table 14.1.
Where that many nuclear hydrogen-generating plants (for example)
might be located is open to question. The coal gasification plants
would be located most economically near the coal resources, and thus
(for example) California's automotive air pollution could be averted
at the expense of creating air pollution in coal resource regions such
as Wyoming's Powder River Basin. Table 14.2 shows a comparison
of the pollutants saved at the vehicle and the pollutants created at a
coal gasification plant with well-controlled emissions [5]. The political implications of this transfer are significant, and some states,
most notably Montana, are already developing policies to the effect
that coal mining in the state may be acceptable but that coal conversion
probably will not be [5]. These state policies are a result of issues
that run deeper than the issue of air pollution. For example, coal
gasification facilities consume vast quantities of water (about 8,200
acre feet per year for 250 million cubic feet per day), and the western
coal-rich regions are water-poor [5]. Even if the coal were brought
to California for gasification, the air pollution in urban areas should
decrease, because pollution can be controlled more readily in a few
large plants than in millions of individual vehicles.

## TABLE 14.2

Comparison of Emissions of Air Pollutants in Gasoline-Powered Cars, Petroleum Refineries, and Hydrogen Production Plants

(all quantities in grams/mile)

| | Carbon Monoxide | Hydrocarbons | Nitrogen Oxides | Particulates | Sulfur Oxides |
|---|---|---|---|---|---|
| Gasoline system | | | | | |
| Vehicle[a] | 14 | 1.6[b] | 2.2 | 0.38[c] | 0.20 |
| Refinery[d] | 0.045 | 0.075 | 0.037 | 0.006 | 0.08 |
| Total | 14 | 1.7 | 2.2 | 0.39 | 0.3 |
| Hydrogen from coal gasification[e] | | | | | |
| Lurgi[f] | — | — | 0.2 | ≪1 | 0.3 |
| Synthane | — | 0.03 | 2.0 | 0.2 | 0.5 |
| Hygas | — | 0.03 | 1.0 | 0.1 | 0.6 |
| Typical | — | 0.03 | 1.0 | 0.1 | 0.5 |

[a]Reference 25. Vehicle emissions are those estimated for the year 1990. However, recent plans to delay (see reference 26) implementation of stricter standards may cause these numbers to be underestimates.

[b]Excludes crankcase and evaporation.

[c]Excludes tire wear.

[d]Using average emission factors for the Los Angeles area refineries (which have the lowest California refinery emissions) and assuming 20 miles per gallon for automobiles (reference 27).

[e]No factors are available for a plant making hydrogen from coal. This should be a good approximation because the emissions from the CO shift reactions needed to produce hydrogen (but not methane) should roughly offset the emissions from the methanation reaction.

[f]Reference 28.

SOCIAL IMPACTS

Safety Considerations

Increased exposure to the safety hazards characteristic of hydrogen is one of the major social impacts that would result from the use
of a hydrogen automobile. Although, like many commonplace things,
hydrogen is a hazardous substance, the hazard to people is not absolute
but is relative to the conditions of hydrogen in actual use. Although
gasoline is also a hazardous substance, society has clearly decided
that the benefits of its use outweigh its hazards. Table 8.1 compared
the safety-related properties of hydrogen, gasoline, and other possible fuels. If stringent safety-related regulations are both promulgated
and followed in the design, production, and use of hydrogen-fueled
automobiles, there is a good possibility that the actual safety hazards
would be no worse (but different in kind) than those accepted in gasoline-fueled automobiles. However, it also seems likely that, at least
initially, individual citizens will not perceive hydrogen to be as safe
as gasoline.

Society as a whole will have a special interest in the possible
dangers presented by a hydrogen-fueled automobile because exposure
will be involuntary once a transition begins. Drivers (even of gasoline-
powered cars), passengers, and even persons passing by on the streets
will be increasingly exposed to hydrogen. State highway patrols,
police, firemen, rescue teams, and ambulance attendants must be
suitably prepared for whatever actual dangers are presented by hydrogen. Individuals within these emergency service groups will require
specialized training to deal with the new hazards posed by hydrogen.
However, since all of these groups normally undergo specialized training, this should not have a major impact.

Individual perceptions of the safety hazards of a hydrogen-fueled
automobile may be very different from the hazards that actually exist,
and therefore individual concern over the safety hazards may assume
the appearance of being exaggerated and unrealistic. As discussed
in Chapter 8, perceptions of "nonreal" hazards are usually based on
incomplete or erroneous knowledge and can also be influenced by sociological and psychological factors. Because of a general lack of
understandable technical information for public consumption, coupled
with conflicting statements, it seems likely that people will mainly
view the safety hazards of a hydrogen car on a "nonreal" basis with
a consequent resistance to a hydrogen-fueled automobile. If a transition were to begin before public acceptance of the hydrogen technology
had occurred in less personally threatening spheres (for example, in
electric utility load leveling), extreme social impact might result from
organized attempts to block or otherwise counter implementation.

Actual safety hazards of a hydrogen-fueled car that must be considered include liquid storage tank ruptures, fuel system malfunctions, vehicle breakdown, and collisions. The most commonly expressed concern regarding hydrogen involves fires or explosions following rupture of a storage tank and fuel spill (even though explosions are exceedingly unlikely). Although less energy is required to ignite hydrogen than gasoline, the ignition temperature required is higher; it has been stated in the literature that while a spark or open flame could ignite hydrogen, a lighted cigarette could not.

Some have asserted that the safety hazards of refueling a hydrogen-fueled automobile would be less than those presented by gasoline refueling [29] because concern with safety and the desire to conserve the boil-off gas would lead to use of a "closed" fueling system. By contrast, a gasoline transfer system is "open" and allows the potentially hazardous (and readily seen) vapors to escape to the atmosphere. Since this assertion is quite reasonable, probably the greatest hazards with hydrogen cars, as in gasoline-powered cars, will come from uncontrolled collisions.

In a collision, there would likely be sparks generated by the impact, the bending and tearing of metal, and the scraping of vehicle parts on the pavement. If spilled hydrogen did not ignite, its high buoyancy and diffusity would lead it to dissipate upward from the scene very rapidly; in contrast, gasoline spills give off vapors that are heavier than air and spread at ground level. If the hydrogen alone were to ignite, its low amount of radiant energy would keep the fire from spreading through the mechanism of heating and igniting nearby objects, but the nonluminous character of the hydrogen flame would make it easier for a person to contact it inadvertently. However, if there were a fire, it is highly unlikely that other objects nearby would escape being ignited. Thus, a hydrogen fire resulting from a collision would still be cause for concern. Metal hydride storage tanks pose a different kind of hazard; while little hydrogen can escape from an unheated ruptured container, the metal hydride powder of some suitable alloys (especially magnesium) can itself ignite and burn.

Special hazards may arise during the transition period when both hydrogen- and gasoline-fueled cars are on the road. A collision involving both types of vehicles might create an especially difficult fire control situation, because a quickly burning hydrogen fire might start a longer, even more hazardous gasoline fire.

## Specialization Required in Production and Maintenance

Actual production of the hydrogen fuel distribution and onboard systems will require more precision than is necessary for a gasoline system because of the exactness required to insure leakproof systems for safety reasons. As a result, some factory workers will need more skills than they need today; similarly, automotive mechanics will require specialized training to detect and rectify problems with hydrogen fuel systems. It is difficult to gauge the magnitude of these impacts at this time because the degrees of specialization required and the overall educational attainments of the population at the time when a transition might begin are difficult to foresee. Regardless of magnitude, most individuals affected will be blue-collar workers.

## Decreased Independence and Self-Reliance

The specialization required to repair the hydrogen fuel system and the dangers posed by a leaky system would probably mean that hydrogen car owners would be more dependent on mechanics for repair work than they are with gasoline-powered cars today. This would affect all hydrogen car owners to some extent, but especially those who prefer to perform some of their own auto repairs.

A number of men customarily have performed their own auto repairs and thereby have both decreased the cost of repair work and demonstrated independence and self-reliance. Many men have taken pride in the mechanical condition of their automobiles, particularly those who have had the mechanical ability to keep their cars in good running condition. Men have been able to exhibit a certain amount of masculine "power" over the world by repairing their own cars. Such individual demonstrations of power seem to be gaining importance as the world becomes increasingly specialized, computerized, and bureaucratic.

The new feminism of the past few years has begun to challenge the "male only" tradition of mechanical ability. Women are becoming more independent and self-reliant in areas that have traditionally been left to men. For example, there are now classes in auto repair specifically intended for women. As a result, women are sometimes seen working on cars. As women increase their mechanical skill, they, too, are being afforded the opportunity to exercise more control over their own lives, giving them new power over the obstacles of the contemporary world. Thus, assuming the continued growth of feminism, women may not relish relinquishing this aspect of their newly acquired independence.

A decrease in independence and self-reliance also has an economically tangible side—the increased cost of automobile upkeep for persons who have in the past performed some of their own auto repairs. For persons unable to afford the cost of repair work, this would probably be seen as a discriminatory feature of hydrogen car ownership.

Many automobile owners show an increasingly resistive attitude toward automobile mechanics, while the greater complexity of automobiles has made it increasingly difficult for the individual to know whether he has been treated honestly. The founding of an agency in California to handle consumer complaints against automobile mechanics and the subsequent large number of complaints received are indicators of this distrust. This attitude is likely to persist and would increase the importance of reduced self-reliance.

An especially important practical consideration will be the decreased ability of owners of hydrogen-fueled cars to handle emergency mechanical breakdown situations, especially those involving the fuel system. This may be a major consideration for individuals who live or travel in rural or remote undeveloped areas.

## Reduction in Air Pollution

The reduction in air pollution derived from the use of hydrogen automobiles is a source of potential major positive impact on health and aesthetics. As air quality has declined, the relationship between air pollution and respiratory ailments has become widely recognized [30]. The benefits of reduced air pollution from hydrogen cars would affect the populace differentially, of course; but those who live or work in large urban areas, where pollution is at its worst, would benefit most. In particular, the poor, the aged, and those in ill health, who are often unable to escape the urban environment, would derive the most health benefits from reduced air pollution.

The relationship between air pollution and aesthetic values has also gained recognition, especially in relation to the deterioration of objects (including works of art), reductions in visibility, and the presence of noxious odors. As with health benefits, society as a whole would be affected by the improved quality of the environment, but the primary benefactors would be individuals living in communities that suffer from extreme air pollution. In addition, some psychological benefits might be derived solely from the knowledge that a step is being taken to improve the environment.

## Independence from Foreign Energy Sources

Having experienced the energy crisis in the winter of 1973-74, Americans now realize more clearly that fossil-fuel resources are limited and that, as a nation, we depend on other countries to supply our increasing demands for liquid fuel. The use of hydrogen fuel derived from nonfossil sources could have major impacts on the sense of security of both the individual and the nation as a whole. Although use of hydrogen as fuel could not quickly reach major proportions, the sense of uneasiness about dependence on other nations for fuel is likely still to be present when the hydrogen era might begin.

Judging from overheard conversations and news reporting, Americans now suffer feelings of anxiety concerning their individual freedom owing to the recent energy "crisis" and the potential for more frequent and larger crises. Many of the deeply held values in our society—geographic mobility, personal freedom, recreation, tourism, prestige, material status, and privacy—are derived in part from private ownership and operation of the automobile. Should the availability of automotive fuel decrease drastically or its price become very high, considerable changes in individual life-styles would follow. These threats are likely to persist and to become even stronger in the era when hydrogen cars might begin to be deployed. The availability of hydrogen as a fuel might enhance feelings of individual freedom by lessening the threat of fuel shortages.

Knowledge that as a nation we were not dependent on other nations for such a basic necessity as fuel might add to a strong national feeling. This is not necessarily beneficial, however, for it could prove detrimental to international relations. It is now recognized that interdependence of nations helps to keep communications open between them. If nations were to become totally independent with respect to such a basic need as energy, and this reduced the need for communication and conciliation, then the possibility of misunderstanding and conflict might increase. In a world of many nations, with conflicting philosophies, interdependence may be an important ingredient in maintaining peace.

### Impacts of Hydrogen Filling Stations

Hydrogen filling stations will probably be larger than their gasoline counterparts [29] because there would be requirements for extra facilities to capture and reliquefy vaporized liquid hydrogen or, perhaps, to store and rejuvenate metal hydride beds. Simply to supply an equivalent amount of energy, a liquid hydrogen station would have to store about three times as many gallons of liquid hydrogen as it would gasoline and would also require proportionally more deliveries. Alternatively, if the hydrogen were liquefied on-site, sizable liquefaction facilities would be required. Moreover, for safety, a larger peripheral buffer of vacant land would probably be required than for a

gasoline station—unless the present standards [29] for hydrogen handling are relaxed (an unlikely prospect).*

The mere size of a hydrogen filling station might become a
source of land use conflict. Besides their larger size, early in the
transition period away from gasoline-powered cars, hydrogen filling
stations would probably be few in number (to ensure operating economies) and probably have a larger capacity than they would in a steady-
state, all-hydrogen automotive system.

These filling stations would have several impacts on citizens re-
siding or working in the neighborhoods where they were located. First,
an abnormally large number of cars would be required to converge on
the few hydrogen filling stations, and this would tend to increase traffic congestion and noise in the vicinity. Additionally, the convergence
of traffic would probably induce changes in neighborhood makeup as
merchants would probably preferentially seek locations nearby because
of the commercial attractiveness of such a strategic point.

A direct impact resulting from the density and placement of hydrogen filling stations, particularly during transition, would be the decreased freedom of mobility of persons operating hydrogen cars. Initially, hydrogen filling stations could not be available in all areas, especially in low-density rural areas. Consequently, the initial hydrogen
car owners may be only those who also have access to a conventional
gasoline-fueled automobile. Thus, ownership of a hydrogen automobile
may initially be limited to families able to afford more than one automobile, with the hydrogen-powered vehicle largely confined to local
use.

Although a hydrogen filling station may prove safer than a gasoline filling station, there would be increased exposure to hazards from
the increased number of hydrogen delivery vehicles on the road. This
would increase the probability of accidents involving delivery vehicles.
Even though the magnitude of this impact is not very great, there are
enough accidents today involving gasoline delivery trucks to warrant
attention to the potential hazard.


ECONOMIC IMPACTS

There were about 101 million automobiles in operation in 1973
in the United States [7]. Of these, about 100 million were operated
privately as personal vehicles [7]. About 9.6 million new cars were

*For example, a 19,000-gallon liquid hydrogen storage tank must
be at least 25 feet from the nearest roadway and 75 feet from the nearest building [29].

sold in 1973, and the replacement rate generally runs about 10 percent per year [7]. These vehicles represent an initial value investment of about $250 billion and have a present value of about $85 billion (although new automobile prices have increased rapidly in the last several years).* Automotive maintenance and ownership contributed about $103 billion in 1973, or 13 percent, to total personal consumption expenditures [13].

Although it is difficult to estimate the total cost of a hydrogen-fueled automobile in the distant future, the cost of automotive hydrogen fuel storage systems can be estimated. For example, it has been estimated that in mass production the cost of liquid $H_2$ containers holding 2 million Btu (the energy equivalent of 16 gallons of gasoline) could be lowered to approximately $1,200 to $1,800 each (compared with $2,000 each in 1972 [32]. Other manufacturers, such as Minnesota Valley Engineering Company, estimate that dewars might cost $300 to $400 when mass produced [33]. Since the technology and manufacturing of small cryogenic dewars is well established, further cost reductions are unlikely without new materials or new concepts. The safety requirements of vehicle storage systems are likely to have a significant impact on dewar costs. Current containers for liquid hydrogen are built for stationary applications, but the need to be able to withstand high-speed impacts from any direction is almost certain to increase cost estimates.

Costs of storage in the form of a metal hydride would depend on the metals employed. The Futures Group estimated that a magnesium hydride system for automobiles would cost about $470 [34], but today there is no large-volume production of metal hydrides on which to base estimates. In recent years, magnesium has fluctuated around $0.38/lb [35];† assuming 500 lbs of magnesium were used per car [12], the cost of the magnesium alone would be approximately $190. Fabrication costs could easily double the cost. A magnesium-nickel hydride ($Mg_2NiH_2$) capable of holding 37 ft$^3$ ($2 \times 10^6$ Btu) of hydrogen would weigh about 1,000 lbs. With magnesium at $0.38 per lb and nickel at $1.50 per lb [36], the material costs alone would be about $1,000. By comparison, an average gasoline tank weighs about 25 lbs and can be manufactured for about $30 [32].

The availability of metals for hydrides will depend on a gradual buildup of consumption. The worldwide resources of magnesium are

---

*Largely because of new air quality, safety, and crash-worthiness regulations.

†Since magnesium production is energy-intensive, these costs can be expected to rise (in constant dollars) to reflect the rising cost of energy.

vast because magnesium salts are abundant in seawater. In the United States, large-scale production of magnesium is concentrated in Utah and Texas. In 1973, the U.S. primary and secondary production of metal was 140,000 short tons, but under wartime conditions, magnesium output was increased from 5,570 tons in 1940 to 155,000 tons in 1943—a level only now being regained [37]. At a use rate of approximately 500 lbs of magnesium per car, 250,000 tons would be needed for every 1 million new cars—or roughly 70 percent more than current domestic magnesium production. Although about 10 million cars are normally produced each year, judging from wartime experience, the implied additional production capability could apparently be brought on line.

Magnesium-nickel hydrides could not be introduced as easily. Roughly 530 lbs of nickel would be required per car. This means that 265,000 tons of nickel would be required for every 1 million cars produced. Consumption of nickel in the United States in 1973 was about 195,000 tons, with nearly all of it (180,000 tons) imported. Canada is currently the main source of U.S. nickel imports, but the future availability of nickel at current prices is doubtful in the light of recent Canadian actions to preserve their resources (oil and gas) for their own use.

The very high cost and weight (with an implied vehicle fuel efficiency penalty) of magnesium-nickel hydride virtually rules out its use in automobiles. Even the high cost of magnesium hydride would lead to some important socioeconomic impacts. First, during a transition when gasoline cars were still available in competition, the implied increased cost of this type of a hydrogen-fueled automobile could discourage buyers from making a purchase. This would particularly affect persons marginally able to afford new cars. Second, the cost of the hydride would carry over to the resale value of the car because the magnesium hydride would retain value as scrap. As a result, prices of used hydrogen cars would also be higher than their gasoline counterparts, and this would have a negative impact on individuals who can only afford low-cost used cars.

A positive effect of the high scrap value of a hydride bed (or a stainless steel liquid hydrogen dewar) would be the increased incentive for vehicle recycling because of the improved economic return to auto dismantlers (normally very marginal businesses). This same residual attractiveness, however, would tend to spread the threat of theft to old cars, which are essentially immune to theft today because of their small residual value.

The investment in the present automobile fuel system may be estimated from the assets of the U.S. petroleum industry by separating the natural gas and chemical production assets of the large integrated companies. Since roughly 50 percent of refinery output is gasoline, it

can be assumed that 50 percent of the value of refineries and supporting facilities can be assigned to the gasoline distribution system. The
net value of production, refining, and distribution assets of the petroleum industry was $50 billion in December 1973 [38]; this is roughly
half the original or gross investment value of $102 billion [38]. The
major components of this investment assigned to the gasoline distribution system are shown in Table 14.3.

To change the present automotive fuel system from gasoline to
hydrogen would require complete replacement of all present facilities.
Hydrogen fuel stations would probably number far fewer than the
218,000 gasoline stations in the United States in January 1973 [39].

Electrical generating capacity and electrolytic hydrogen production sufficient for 100 million cars would require deployment of about
540 1-GW nuclear power plants (for example), and, although there is
debate about the future cost of nuclear electric plants, at $560 per
kW (1973 dollars)* the total electric plant investment would be about
$300 billion (1973 dollars). The electrolysis facilities could cost
about $18 billion (1973 dollars) and liquefaction facilities could cost
about $20 billion (1973 dollars).† Thus, the total investment would
be about $340 billion (1973 dollars). Because there will be many com-
peting uses for investment funds of this magnitude, such investment
would have to be spread over a long period of time.

The investment needed for hydrogen distribution and marketing
facilities would be in addition to the $340 billion already discussed.


TABLE 14.3

Estimated Investment in U.S. Gasoline Distribution System—
December 1973
(in billions of dollars)

| | |
|---|---:|
| Production assets | 14.4 |
| Transportation and distribution | 22.3 |
| Refineries | 3.6 |
| Marketing and administrative | 7.5 |
| Total | 47.8 |

Source: Reference 38.

---

*See Chapter 16 for a discussion of nuclear power plant investment
costs.

†See Chapter 11 for a discussion of these investment costs.

The future cost of a hydrogen fuel station will obviously depend on the nature of the facility; if it is analogous to today's service station, then it will provide repair and maintenance services as well as fuel. It will quite likely be larger and more costly than present gasoline stations because of more costly storage and transfer facilities. Currently, the average cost for new gasoline stations is about $250,000 for land and buildings [40]; at this rate, a low estimate for 100,000 new hydrogen filling stations would require $25 billion.

The total investment in electrical generating plants and electrolysis and liquefaction plants could easily exceed $250 billion for 100 million vehicles. This compares to a total investment of about $113 billion for the entire U.S. petroleum industry (including natural gas and petrochemical facilities) in 1973 [38].

The gasoline fuel distribution system employed about 1 million persons in 1973 [7]. Although it is very difficult at this time to estimate the employment needs of individual components of a potential liquid hydrogen (for example) automotive fuel system, it is reasonable to expect that it would employ about the same number of people.

In 1973 consumers paid $4.7 billion (net of premiums less claims paid) to the automobile insurance industry [7]. This industry would surely become involved in the question of the relative safety of hydrogen vehicles. The recent trend, illustrated by the public issues of crashworthy bumpers, air bags, and no-fault insurance, is for automobile insurance companies to try to influence both public and congressional attitudes through advertising and lobbying. As discussed in Chapter 8, the issue of hydrogen safety is too complex to foresee now whether a hydrogen car would warrant the opposition or support of the insurance industry in advance of actual large-scale experience.


### RESOURCE UTILIZATION

Transition to a hydrogen-fueled automotive system would not be the most efficient use of U.S. energy resources. As Figure 7.7 showed, the net cascaded system efficiency of hydrogen automotive systems is inferior to other synthetic fuel alternatives based on the same resources—even allowing for the claimed superior efficiency of hydrogen combustion in engines. Figure 14.3 extends the comparison further by comparing the coal- and nuclear-based hydrogen systems shown in Figure 7.7 with coal- and nuclear-based electric car systems employing the still speculative technology of lithium sulfur batteries [41]. Figure 14.3 demonstrates that a transition to a hydrogen-fueled automotive system is less compatible with a future in which stringent energy conservation would place a premium on optimum use of basic energy resources than would be a transition to advanced electric cars.

FIGURE 14.3

Energy Resource Utilization of Hydrogen-Fueled and Electric Cars

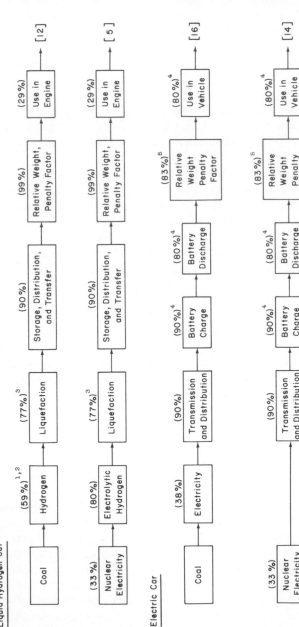

Notes:

1. Includes energy utilization for coal to product, coal for process heat, and coal for steam and electricity.

2. SRI internal data.

3. Reference 38.

4. Reference 41.

5. Assumes battery and ancillary equipment weight of about 1,5000 lbs.

Because the electric car would probably be even more beneficial for urban air quality than the hydrogen car (no nitrogen oxides but a small amount of ozone),* considerable public debate over the relative merits of these options must be expected before either one is chosen. Unless, however, advanced lithium sulfur (or the equivalent) high-energy and high-power density batteries become practical, the future of the electric car is not bright. Just as considerable discussion over the safety aspects of hydrogen automobiles is to be expected, so there would be debate about any battery involving molten† alkali metals (such as lithium or sodium), because alkali metals are very dangerous substances when they contact moisture or water, and nearly all metals are dangerously hot when molten.

## FLEET VEHICLES

### Decisions to Deploy

Because fleet vehicles require so much less in the way of a fuel distribution network, they can be more readily deployed and supported. Accordingly, the implementation scenarios of Chapter 13 show hydrogen-fueled fleet vehicles coming into use sooner than their private automobile counterparts. It is extremely significant that fleet operators normally provide the fuel for their own vehicles by means of fleet accounts with oil companies [3], thereby eliminating the need for a series of public filling stations. The prospect and possible guarantee of a bulk sale to a fleet operator would be more attractive to suppliers than sales on the open market to many individual owners [3]. Consequently, the three-way dilemma mentioned earlier for private automobiles is more readily resolvable in the fleet vehicle context because the decision of a single fleet owner can affect the sale of a large number of vehicles and their fuel. Moreover, fleet operators are mainly interested in the economics of purchase, maintenance, and operation rather than in the less tangible psychological aspects of car ownership. Thus, when a convincing case in favor of hydrogen-fueled cars or vans can be made on economic grounds, even a few favorable fleet operator decisions could create a market large enough to initiate commercializa-

---

*Ozone is a powerful oxidizing agent and is a key oxidant product of photochemical smog—reaction products of unburned hydrocarbons, nitrogen oxides, and sunlight [30].

†The lithium sulfur battery has an operating temperature of about 400°C (750°F) [30].

tion.  For example, a few decision makers in the New York City or
Washington, D.C., taxi business could lead to complete use of hydrogen-fueled taxis in those cities, and this would encompass hundreds
of cars and commercially significant amounts of fuel.

It is also very important that some of the most significant fleet
operators are the federal, county, and municipal governments that
have a wider range of concerns and objectives than either individual
car owners or commercial fleet operators.  Governments could decide
to implement hydrogen as a fuel for their fleet vehicles (even at higher
cost) solely on the basis of providing a "good example" for the improvement of air quality.  Today such municipal decisions are, in fact,
being made for propane-powered cars and small buses.  Some electric
vehicle makers are seeking to establish a beachhead by recourse to
such decision making.  The most notable example is the trial of a small
fleet of electric-powered postal delivery vans by the U.S. Post Office
[42].  If successful, the 240,000 light-duty postal vehicles would
present a substantial initial market.

## Impacts

By far, the single most important impact stemming from experience with hydrogen-powered fleet vehicles would be the psychological
and educational conditioning effect on the public and the establishment
of a rudimentary hydrogen fuel distribution network that could be expanded into a more general system for general automotive use.  Successful demonstration of safe, economical, and reliable operation of
hydrogen-powered urban buses would probably yield the most conditioning of the general public—either for or against hydrogen.  However,
because buses are commonplace, readily identified, and entrusted
with the simultaneous safety of many people, the question of hydrogen's
actual (or perceived) safety will be particularly salient.

It is likely, however, that public reticence about hydrogen-
powered buses will be pronounced, and their introduction may require
prior successful experience in privately owned fleet vehicles.  Municipal governments, labor unions (especially for drivers), the riding public, and the general public (each of whom might be involuntarily exposed
as a bystander) must all be convinced that a hydrogen-powered bus is
safe enough to warrant a try.  Municipal officials, with ultimate responsibility to the public through elections, can be expected to be
cautious in embracing a new technology if a significant number of constituents express concerns over safety.  Labor unions, because of
their ability to articulate the collective concerns of their members,

## TABLE 14.4

## Societal Impacts of a Transition to Hydrogen-Fueled Private Automobiles and Fleet Vehicles

| Impact Class | Stakeholders | Nature of Effect | Magnitude of Impact (units) |
|---|---|---|---|
| | | Private Automobiles | |
| **Economic** | | | |
| Investment required for production and distribution of hydrogen fuel | Oil companies, Utilities, Capital market | Large demand for capital to supplant present system | Major ($) |
| Investment required to produce new class of vehicles | Auto manufacturers, Component manufacturers, Cryogenic industry | Large demand for capital to supplant present system; Development of new markets | Major ($) |
| Cost of vehicles, fuel, maintenance, and repair | Consumers | Access to vehicles may be restricted by higher costs | Major ($, people) |
| **Institutional** | | | |
| Insurance | Insurance companies, Consumers | Effect of safety experience on insurability | Moderate ($, people) |
| Ownership of fuel distribution system | Public energy utilities, Oil companies, Pipeline companies, Gas producers | Competition for control of hydrogen fuel markets | Major (power) |
| Regulatory | Hydrogen fuel distributors, Government regulatory bodies, Hardware producers | Expanded regulatory authority over synthetic gas and subsequent uses of hydrogen; Safety and performance criteria | Major (power) |
| **Labor force** | | | |
| Fuel distribution | Dealer and employees, Unions, Customer | Specialized training to handle more exotic fuel and materials | Moderate (people, $) |
| Vehicle maintenance and repair | Mechanics, Unions | Specialized training to deal with more sophisticated systems | Moderate (people) |
| Industrial shifts | Factory workers | Threat to employment as industry changes | Minor (people, $) |

(continued)

TABLE 14.4 (continued)

| Impact Class | Stakeholders | Nature of Effect | Magnitude of Impact (units) |
|---|---|---|---|
| Individual values<br>Personal safety | Vehicle passengers<br>Service personnel<br>Emergency crews<br>Bystanders | Real or perceived threat to personal safety by exposure to hydrogen | Major (people) |
| Independence/self-reliance | Vehicle owners | Increased reliance on specialized station attendants, mechanics<br>Altered freedom of mobility | Moderate (people) |
| Health | Vehicle users<br>Urban residents<br>Governments | Improved air quality affecting health | Minor to major as transition proceeds (people, $) |
| Aesthetics | Urban residents | Decreased deterioration of objects by air pollution<br>Increased visibility<br>Decreased noxious odors | Minor to major as transition proceeds (people, $) |
| Sense of personal freedom | Individuals | Lessened fear of interruption of life-style caused by actions of foreign governments regarding energy supplies | Minor (people) |
| National security | U.S. citizens | Independence from foreign control of fossil-fuel resources | Major (people) |
| Resource utilization | Society | Suboptimum use of basic energy resource compared to alternatives | Very major ($) |
| Land use | Nearby residents | Increased size of filling stations and traffic density, changing land use in vicinity | Minor (people) |
| | | Fleet Vehicles | |
| Safety | Labor unions | Perceived or actual hazards associated with use of hydrogen | Moderate (people, $, power) |

would also play a significant role in decisions to deploy hydrogen-powered buses.*

Many of the impacts to be expected for use of hydrogen in private vehicles would be unimportant for fleet vehicles or lesser in magnitude. In particular:

1. Decreased personal ability to repair the fuel system is basically irrelevant because the fleet owner rather than the driver normally provides the repair service.

2. Safety education would be facilitated by incorporation into existing operator training programs.

3. Air quality improvement potential would be less because fleet vehicles account for about 30 percent of all mileage in the automotive sector.

4. The effect of increased initial cost would be less because fleet owners use business accounting methods that allow capital recovery through depreciation.

5. Fleet-owned filling stations are privately and exclusively operated, so merchants would not be attracted to the vicinity, thereby lessening the secondary land use impact of filling stations.

## SUMMARY

In general, use of hydrogen in fleet vehicles can be expected to precede use in private vehicles and to establish the general attitude of the public toward hydrogen-powered automotive vehicles. Table 14.4 summarizes the impacts that would be expected from automotive use of hydrogen.

## REFERENCES

1. U.S., Senate, Committee on Public Works, History and Future of Spark Ignition Engines, report prepared by the Environmental Policy Division of the Congressional Research Service, Library of Congress, at the request of Senator Edmund S. Muskie, 93rd Cong., 1st sess., September 1973.

2. "Gasoline: EPA Issues Rules, Proposal on Availability of Unleaded Gasoline," Energy Users Report, May 9, 1974, pp. C-2 and C-3.

---

*Labor unions could be expected to have a voice in nearly every use of hydrogen in fleet vehicles.

3. F. H. Kant et al., "Feasibility Study of Alternative Fuels for Automotive Transportation," vol. I: Executive Summary; vol. II: Technical Section; vol. III: Appendices, Exxon Research and Engineering Company, Linden, N.J., for the U.S. Environmental Protection Agency, June 1974.

4. J. Pangborn and J. Gillis, "Alternative Fuels for Automotive Transportation—A Feasibility Study," vol. I: Executive Summary; vol. II: Technical Section; vol. III: Appendices, Institute of Gas Technology, Chicago, Ill., for the U.S. Environmental Protection Agency, June 1974.

5. E. Dickson et al., "Impacts of Synthetic Liquid Fuel Development for the Automotive Market," Stanford Research Institute, Menlo Park, Calif., for the Energy Research and Development Administration, Washington, D.C. (in preparation).

6. J. E. Johnson, "An Economic Perspective on Hydrogen Fuel," in Hydrogen Energy, ed. T. N. Veziroglu (New York: Plenum, 1975), pp. 299-308.

7. Statistical Abstract of the United States, 1974 (Washington, D.C.: Bureau of the Census, U.S. Department of Commerce, July 1974).

8. R. E. Billings and F. E. Lynch, "Performance and Nitric Oxide Control Parameters of the Hydrogen Engine," report no. 73002, Billings Energy Research Corporation, Provo, Utah, April 1973.

9. J. G. Finegold et al., "The UCLA Hydrogen Car: Design, Construction and Performance," paper no. 730507 presented at the Society of Automotive Engineers, Automobile Engineering Meeting, Detroit, Mich., May 14-18, 1973.

10. J. G. Finegold and W. D. Van Vorst, "Engine Performance with Gasoline and Hydrogen: A Comparative Study," in Hydrogen Energy, ed. T. N. Veziroglu (New York: Plenum, 1975), pp. 685-96.

11. R. R. Adt et al., "The Hydrogen-Air Fueled Automobile Engine," Eighth Intersociety Energy Conversion Engineering Conference, 1973, pp. 194-97.

12. A. L. Austin, "A Survey of Hydrogen's Potential as a Vehicular Fuel," UCRL-51228, University of California Radiation Laboratory, Livermore, Calif., June 1972.

13. R. J. Schoeppel, "Design Criteria for Hydrogen Burning Engines," final report, EPA Contract EHS-70-103, October 1971.

14. P. B. Dieges et al., "An Answer to the Automotive Air Pollution Problem," First Annual Report, Perris Smogless Automobile Association, Perris, Calif. (undated).

15. P. Underwood and P. B. Dieges, "Hydrogen and Oxygen Combustion for Pollution Free Operation of Existing Standard Automotive Engines," Sixth Intersociety Energy Conversion Engineering Conference, 1971, pp. 317-22.

16. "Fuel of the Future," Time, September 11, 1972, p. 46.

17. L. Lessing, "The Coming Hydrogen Economy," Fortune, November 1972, pp. 138-46.

18. "When Hydrogen Becomes the World's Chief Fuel," Business Week, September 23, 1972, pp. 98-102.

19. W. Clark, "Hydrogen May Emerge as the Master Fuel to Power a Clean Air Future," Smithsonian, August 1972, pp. 13-18.

20. P. Gwynne, "The Hydrogen Car," New Scientist, October 18, 1973, pp. 202-03.

21. R. Breshears, H. Cotrill, and J. Rupe, "Partial Hydrogen Injection into Internal Combustion Engines, Effect on Emissions and Fuel Economy," presented at the Environmental Protection Agency, First Symposium on Low Pollution Power System Development, Ann Arbor, Mich., October 14-19, 1973.

22. W. J. D. Escher, "Survey and Assessment of Contemporary U.S. Hydrogen-Fueled Internal Combustion Engine Projects," Tenth Intersociety Energy Conversion Engineering Conference, Newark, Dela., August 18-22, 1972, pp. 1143-55.

23. A. Hammond, "Fission: The Pros and Cons of Nuclear Power," Science 178, no. 4057 (October 13, 1972): 147-49.

24. M. Willrich and T. B. Taylor, Nuclear Theft: Risks and Safeguards (Cambridge, Mass.: Ballinger, 1974).

25. "Compilation of Air Pollutant Emissions Factors," U.S. Environmental Protection Agency, 1973.

26. "[President] Ford Asks 5-Year Freeze on Auto Emissions Curbs," New York Times, June 28, 1975, p. 42M.

27. "Environmental Considerations in Future Energy Growth," vol. I, Battelle Memorial Research Institute for the Office of Research and Development, Environmental Protection Agency, Contract no. 68-01-0470, April 1973.

28. "Draft Environmental Statement for the El Paso Gasification Project, San Juan County, New Mexico," Upper Colorado Region, Bureau of Reclamation, U.S. Department of the Interior, July 16, 1974.

29. W. E. Stewart and F. J. Edeskuty, "Logistics, Economics, and Safety of a Liquid Hydrogen System for Automotive Transportation," paper presented at the Intersociety Conference on Transportation, Denver, Colo., September 23-27, 1973. American Society of Mechanical Engineers Publication 73-ICT-78.

30. S. J. Williamson, Fundamentals of Air Pollution (Reading, Mass.: Addison-Wesley, 1973).

31. U.S., Department of Commerce, Bureau of Economic Analysis, Social and Economic Statistics Administration, Survey of Current Business, July 1974, p. 24.

32. J. E. Johnson, "The Storage and Transportation of Synthetic Fuels," a report to the Synthetic Fuels Panel working on the Federal Council on Science and Technology R&D Goals Study, September 1972.

33. William J. D. Escher, Escher Technology Associates, St. John's Mich., personal communication, 1973.

34. E. Fein, "A Hydrogen Based Energy Economy," The Futures Group, Glastonbury, Conn., October 1972.

35. "Commodity Data Summaries, 1974, Appendix I to Mining and Minerals Policy," Bureau of Mines, U.S. Department of the Interior.

36. J. Paone, "Magnesium," in Mineral Facts and Problems, Bureau of Mines Bulletin 650, 1970.

37. "Capital Investment of the World Petroleum Industry, 1973," Chase Manhattan Bank, New York, December 1974.

38. "The Vacancies on Gasoline Alley," Business Week, December 15, 1973, pp. 20-21.

39. National Petroleum News, Factbook Issue, 1974 (March 1974), p. 58.

40. P. A. Nelson et al., "The Need for Development of High Energy Batteries for Electric Automobiles," Argonne National Laboratory Report ANL 8075, November 1974.

41. "When Electric Trucks Deliver the Mail," Business Week (Industrial Edition), March 9, 1974, pp. 78C-78F.

42. W. R. Parrish and R. O. Voth, "Cost and Availability of Hydrogen," in Selected Topics on Hydrogen Fuel, ed. J. Hord, Report NBC IR 75-803, Cryogenics Division, National Bureau of Standards, Boulder, Colo., January 1975.

# 15

## COMMERCIAL AVIATION

### THE SYSTEM AND THE STAKEHOLDERS

Commercial aviation is a complex, capital- and energy-intensive system. Since World War II, the United States has been the leader in aviation technology, the production of aircraft, and commonplace air travel.* The chief stakeholders in the U.S. commercial aviation sector are:

Airlines
Aircraft manufacturers
   Airframe
   Engines
Airport operators
Fuel producers and distributors
Government regulatory agencies
   Markets and fares (Civil Aeronautics Board [CAB])
   Safety, air traffic control, and navigation (Federal Aviation
    Agency [FAA])
   Noise (EPA and FAA)
   Fuel use (Federal Energy Administration [FEA])
Consumers

Transition to a hydrogen-fueled aviation system in the United States would affect all these stakeholders as well as their counterparts in

---

*In the non-Communist portion of the world, about 83 percent of the commercial jet aviation is U.S.-built, and 43 percent is U.S.-operated.

foreign nations—especially since the United States has been the dominant supplier of commercial aircraft.

As discussed elsewhere in this book, the fuel distribution network is a key element in any transportation system. Clearly, a hydrogen-fueled airplane cannot operate without assured distribution of fuel.
In strong contrast, however, to the automotive sector, where the final
link in the fuel distribution network is manifested by small, nearly ubiquitous filling stations, * just a few airports account for most of the
fuel dispensed in commercial aviation. The top 10 aviation hubs † accounted for 70 percent of all the domestic plus international enplanements in 1968 [1]. To illustrate, Figure 15.1 shows the top 23 air
traffic hubs in the contiguous United States [2]. Thus, if hydrogen-
fueled commercial aircraft were deployed, a very large portion of the
commercial air transportation system could be fueled from only a few
major airports. Moreover, the low number of major air traffic hubs
is a parallel to the favorable situation found in fleet automotive application where relatively few decision makers control the outcome.

## The Airlines

As potential operators of hydrogen-fueled aircraft, the domestic
U.S. trunk airlines have an especially important stake. As shown in
Figure 15.2, new generations of airplanes have been introduced at
10- to 12-year intervals [3]. Each succeeding generation offered important advances in speed, capacity, and passenger comfort; and adoption of each was essential to enable individual airlines to attract and
hold passengers, because competition had become increasingly a matter of service and cost reduction rather than ticket price differentials.
Indeed, ticket prices have been regulated by the government to such
an extent that except for special restricted fares** the basic rates
have been identical. Service differential aspects of competition have
been manifested in passenger amenities (such as meals, seat and leg-
room, alcoholic drinks, inflight movies, and so on), solicitous ser-

---

*In 1974 there were about 215,000 gasoline filling stations in the
United States.

†A "hub" may consist of several airports such as Kennedy, La
Guardia, and Newark in the New York hub. The 10 hubs with the most
enplanements in 1968 were (in descending order): New York, Los Angeles, Chicago, San Francisco, Atlanta, Miami, Washington, Boston,
Dallas, and Honolulu.

**With stipulations on length of stay, age of the traveler, familial
relationship of passengers, and so on.

FIGURE 15.1

Major Air Traffic Hubs, 1973

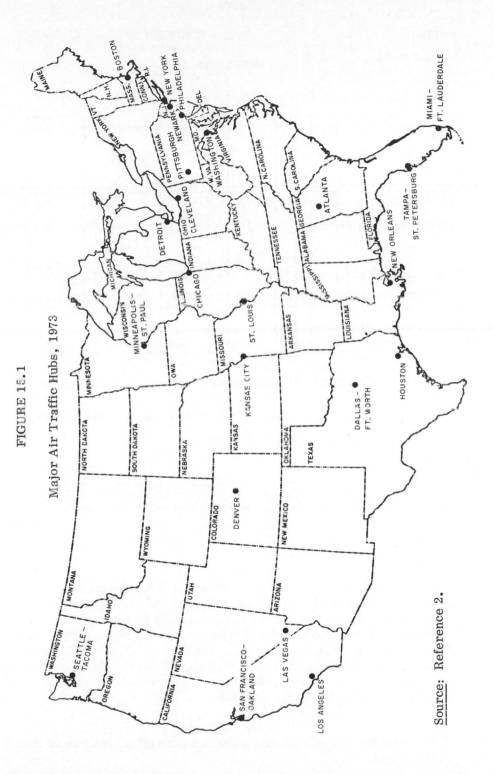

Source: Reference 2.

FIGURE 15.2

U.S. Domestic Trunks—Scheduled Service, 48-State Basis

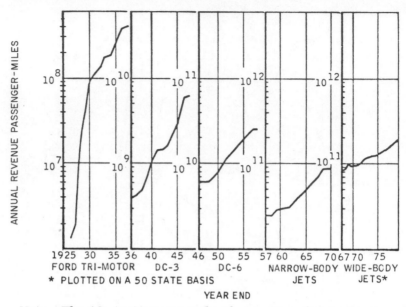

Note: The 10- to 12-year cycle of new aircraft introductions is clearly shown.
Source: Reference 3.

vice, schedule frequency, and low load factors.* But even these distinctions have tended to even out, because within a few years after the introduction of new aircraft, airlines not only fly basically the same equipment (with only minor variations in interior appointments) but they also closely emulate each other in service.

The newest generation of commercial aircraft—the wide-body jets—appear to be nearing the feasible limits of passenger comfort amenities. Meanwhile, airlines are becoming increasingly pressed for funds for capital investment and operating costs, especially for fuel. As a result, it is widely assumed in the industry that the next generation of aircraft will have to offer airlines significant operational savings to gain acceptance [4–8]. Airlines will be especially interested in the following factors for future aircraft:

---

*The fraction of available seats occupied by passengers on any given flight.

1. Initial cost
2. Energy utilization
3. Availability of specified fuel
4. Cost of specified fuel
5. Maintenance and other operating costs
6. Efficiency of aircraft turnaround on the ground
7. Labor utilization
8. Environmental acceptability

In addition, increasingly stringent noise abatement regulations are adding to the financial concern of airlines [5] because of the high cost of retrofitting engines to make them quieter and the effect of airport curfews on scheduling. Thus, under the pressure of governmental noise regulations, airlines will also be seeking much improved noise performance in new aircraft.

## Aircraft Manufacturers

The makers of finished airplanes and engines normally work closely with the airlines during the design of new aircraft. Because of this interaction, new designs for airplanes generally reflect advances in technology and the airlines' desires [9]. Not all airlines have an equal voice, however, in any given design. Those operating the largest or traditionally the most modern fleets are naturally the most courted potential buyers. Thus, the basic aircraft finally offered for sale may especially reflect the needs of one or just a few airlines. *

The development of new aircraft entails substantial economic risks. Because of the enormous development costs of present-day aircraft, the cost per airplane is sensitive to the (small) number produced and sold. Development costs have exceeded $1 billion for each of the most recent generation of wide-body airplanes [1]. At a sales price of $25 million, 400 aircraft sales are needed to bring the average development costs down to 10 percent of the sales price (that is, $2.5 million), a reasonable goal. To show a profit on a new airplane,

---

*For example, the Lockheed Electra design largely reflected the needs of Eastern Airlines for making landings at the relatively difficult La Guardia Airport in New York and (for the era) carried more sophisticated electronic gear than many other airlines really desired. The design of the Boeing 747 was slanted toward Pan American's anticipated needs on long-range, transoceanic routes, and as a result has proved less satisfactory on the shorter domestic routes of other airlines [9].

an airframe manufacturer must assess the size of the market accurately and spread development costs appropriately. If a sales goal of 400 planes is missed by 100 planes, then the company must absorb 25 percent of the development costs. It is common knowledge [10] that it will be many years before the wide-body jets reach the sales goals of their makers.*

Besides the risks involved in estimating the size of the market, manufacturers have a major problem of cash flow associated with the long lead times, large development costs, and large expenditures to get into production. When ten years elapse between the design go-ahead and the date of first delivery, even when followed by reasonable production rates, cumulative cash flow can be negative for 15 years. At the end of 1972 the net worth of each of the three major air transport manufacturers was under $1 billion, as shown in Table 15.1. Thus, the commitment to develop a new generation of planes involves a decision to risk the entire stockholder equity, a process that has been termed "betting the company" [1].

An important factor affecting manufacturers' decisions to embark on development of hydrogen-fueled aircraft is the future policy of the Department of Defense (DoD). Historically, technological advance in the aviation industry has been spawned by DoD needs for ever more sophisticated air weapons and support systems. This technology has been very successfully transferred within the industry itself to civilian aircraft. At present, however, DoD exhibits little interest in hydrogen-fueled aircraft, probably because of worldwide logistics problems with hydrogen and the difficulty of designing a high-performance hydrogen-fueled fighter [11]. It would be awkward to fuel jet transports on hydrogen, but jet fighters on conventional fuel, because

TABLE 15.1

Net Worth of Airframe Manufacturers, 1972

| Company | Net Worth (millions of dollars) |
|---|---|
| Boeing | 865 |
| McDonnell Douglas | 834 |
| Lockheed | 266 |

*Lockheed reportedly is charging research and development costs of the L-1011 against minimum sales of 300 airplanes. In mid-1975, 143 more sales are needed to meet that goal [10].

that would require a parallel fuel system. Should the DoD decide not to fly hydrogen-fueled aircraft for any mission, then the manufacturers' cost of development of a civilian airplane would rise, since there would be no appreciable technology transfer. This would be an important consequence for the future of hydrogen-fueled civilian aviation.

## Airport Operators

Major airports in the United States are publicly owned and operated by municipal or regional governments with the exception of National and Dulles in and near Washington, D.C., which are operated by the FAA. Local governments usually regard operation of a modern, efficient airport as a necessity to retain (and attract) industry and commerce in the region. Airport construction is usually financed through the issue of revenue bonds, backed in part by the anticipation of landing fee receipts collected from the airlines.

Fuel is usually dispensed at airports by one or a few suppliers bound by contract to the airport operator, who acts as a fuel broker with the airlines. Since the investment in fuel-dispensing facilities is large, the fuel suppliers usually obtain "evergreen" contracts— that is, contracts that are normally renewable provided performance is adequate. Airports therefore have a clear interest in the fuel used by the aircraft, because this affects their relationships with both the airlines and fuel suppliers, and affects their physical plant as well. In addition, airports must be concerned about the personal safety of airport personnel, travelers, and the security of the expensive aircraft crowded together.

Airport operators, to a certain extent at least, are responsive to the concerns of the neighboring public—especially with respect to noise. Increasingly outspoken criticism of airport noise by residents and businesses near airports [12, 13] has led airport operators to issue noise abatement operating procedures. These dictate aircraft operations during taxiing and pre-takeoff engine run-up as well as during and just after takeoff. In some cities major lawsuits by airport neighbors against airport operators have led to the purchase and demolition of nearby residential areas to create noise buffer zones around the airport (Los Angeles International is a good example of this). Alternatively, some airports, such as National Airport in Washington, D.C., have established night and morning curfews on airport operations to lessen aircraft noise problems. These actions by airport operators illustrate the pressure airports can exert on airlines and fuel suppliers, through which they can have a major influence on certain aspects of aircraft operations.

## Fuel Producers and Distributors

Fuel producers and distributors naturally play a major role in the aircraft system configuration. As long as the fuel remains unchanged—as it has for the last two generations of aircraft—their role remains rather passive, but should a shift in fuel become a serious possibility, their role will become active and they will constitute perhaps the most important decision element in the system.

As noted in Table 7.1, the weight of the liquid hydrogen fuel carried by a 400-passenger subsonic passenger aircraft with a range of 3,000 nautical miles would be about 28,000 lbs—or 14 tons [14]. This amounts to about one-quarter of the daily output of the largest (60-ton) liquid hydrogen facility ever built. As Table 12.1 shows, the projected hydrogen liquefaction building block of 250 tons per day would be able to fuel about 18 aircraft per day and would use about 60 percent of the daily output of a nuclear/electrolysis building block.

Because it is not economically feasible to convert a commercial* airplane from conventional jet fuel to hydrogen [15], whatever fuel is chosen for the next generation of aircraft will have to be in reasonably certain supply for the lifetime of the airplane. Since design and certification of a new airplane takes 6 to 10 years, and the airplane generally remains in service for about 20 years, the fuel chosen must be available for about 20 years following the delivery of the last airplane in the production run. Thus, if in 1985 a firm decision is made for delivery in 1995 of a new generation of airplanes that uses conventional petroleum-based jet fuel, and if the airplane remains in production for 10 years (until 2005), then jet fuel must remain available until about 2025. In other words, a 1985 decision must be viable for about 40 years. This exceeds the normal planning horizon of most corporations.

As domestic petroleum supplies become increasingly scarce, airlines will naturally want assurance that they will continue to be able to obtain fuel before making a commitment to another generation of petroleum-based conventional jet-fueled aircraft. Moreover, the airlines are unlikely to forget that during the Arab oil embargo in the winter of 1973-74 they received disproportionately small fuel allocations, apparently because the government judged that air travel was less essential †to national welfare than other uses of petroleum—including automobiles. As a result, airlines and aircraft manufacturers are just beginning to give some attention to the aircraft fuel of the future.

In spite of their experience during the Arab oil embargo, elements of the aviation industry often articulate the point of view that the

---

*As distinct from a unique demonstration airplane.
†Or politically less controversial.

absolute need for a portable high-energy density fuel is so nearly
unique to aviation that aviation should be given the highest priority for
petroleum fuels.  However, since they recognize that such priority may
not be politically feasible, their next favored alternative is the production of synthetic jet fuel from coal or oil shale, because this places
all of the burden of accomplishing the change outside the aviation industry onto the shoulders of the petroleum companies.  Clearly,
therefore, fuel producers and distributors will play a central role in
any possible switch of aviation over to hydrogen.


## Government Regulatory Agencies

In general, the role of the FAA would be expected to be more
direct and to begin at an earlier date than that of the CAB.  The primary responsibilities of the FAA (established in 1958) are the establishment and maintenance of U.S. airways (including en route, terminal,
and final approach navigational aids), airport development (a program
of financial assistance), pilot and aircraft certification, setting and
enforcing safety standards for manufacture and operation, and research
and development (mainly related to safety, such as for all-weather in-
strument landings) [16].

The first involvement of the FAA in the airworthiness certification of new aircraft occurs a month or so after the first sales are
concluded and it is decided to go ahead with detailed development of
the conceptual design [17].  Since the first sales are based on only
tentative designs, the FAA participates from the start in assuring that
safety standards are met.  The aircraft manufacturers give great attention to this interaction with the FAA because the FAA must certify
that an airplane class is airworthy before deliveries can be made to
airlines [17].

The manufacturers are guided by existing Federal Aviation
Regulations (FARs), but if a particular situation is not covered because
a novel approach or technology is being planned, the FAA will issue
"special conditions" as a supplement to the relevant FARs [17].  Regulations are couched in terms of a functional capability.  To meet special conditions, the burden of proof is on the manufacturer; and on its
own volition or at the request of the FAA, there may be special experimental verifications of any new technology before it can be incorporated
into the design [17].  Because the stakes are high, the relationship between the manufacturers and FAA representatives is very close during
the entire design period.  Indeed, the FAA representatives are permanently assigned to a particular manufacturer and have offices on
the premises.  Likewise, manufacturers have special staff assigned
permanently to the task of providing FAA liaison [17].

In the past, the aviation state of the art has been mainly advanced by military development* and proved by military experience; this has been reflected not only in the manufacturers' design approach but in the airworthiness standards as well. However, a hydrogen-fueled commercial aircraft may not have been preceded by military experience. Thus, whether the design state of the art will represent levels of safety comparable to those achieved by a similar hydrocarbon-fueled aircraft is uncertain. A further, and perhaps more important, question is whether the aircraft certification process would accept the existing technology or would require that it be advanced.

Of course the actual safety of an aircraft is always demonstrated in practice. The many factors that contribute to accidents, such as pilot proficiency, weather conditions, aircraft maintenance, operation and availability of navigation aids, and aircraft design limitations or malfunctions, are investigated by the National Transportation Safety Board (NTSB). Since circumstances often interact to cause an accident, an unequivocal determination of design inadequacy is not always possible. Nevertheless, the operating experience of an aircraft feeds back in an important way to affect design modifications. The most serious of the original design inadequacies result in an airworthiness directive to owners that specifies a mandatory retrofitted change. An aircraft may even be withdrawn from service until the modification is made.

There is no numerical standard for aircraft safety in terms of injuries or fatalities per million passenger miles. However, whenever a series of accidents occurs, congressional hearings usually follow quickly. These hearings can be considered a bellwether of public opinion and, together with the NTSB crash investigations, have usually led to changes of practice or design that have resulted in acceptably safe aircraft.

Several alternative strategies can be defined regarding the safety of hydrogen-fueled aircraft:

1. Develop a state-of-the-art aircraft and depend on the initial operating period to provide actuarial data on safety and feedback on design deficiencies and operational limitations.

2. Require the manufacturers and/or airlines to engage in an extended period of preoperational (prepassenger) analysis and flight testing. This might be in the form of all-cargo operations that use only airports distant from population concentrations.

3. Fund a government program of research, development, and operational testing, including NASA, DoD, FAA, and airframe and engine manufacturers.

---

*With some contributions from NASA in close coordination with DoD.

Basically, these alternatives represent a trade-off between a program of in-service development and testing involving the traveling public and a program of extensive preoperational development and testing. This amounts to a choice between private funding, ultimately passed on to the traveling and shipping public, and direct public funding, putting the burden on taxpayers at large.

Crash-worthiness or passenger survivability is becoming an increasingly important aircraft safety issue. In many aircraft accidents, the occupants would have survived if there had been no fire. Therefore, a primary concern in crash-worthiness is with postcrash fires, including those involving the fuel supply. Even if hydrogen were not flammable and explosive within a wide range of hydrogen-to-air ratios, liquid hydrogen would be a dangerous material because of its extreme coldness. Concern would be warranted to ensure that an accident did not spill hydrogen on people, or so thoroughly chill an area that evacuation would be endangered.

The close working relationship between the FAA and the manufacturers essentially ensures that neither party is surprised when the final airworthiness flight tests occur [17]. The close relationship, however, is subject to the criticism often raised against regulatory agencies—there is a significant danger that the welfare of the manufacturer will be given more weight than the welfare of the public. Nevertheless, the effectiveness of safety regulation is demonstrated by the declining death rate per distance traveled, shown in Table 15.2, in spite of increasing traffic and airport congestion.

TABLE 15.2

Domestic Air Travel by Scheduled Air Carriers: Statistics

| Category | 1950 | 1960 | 1970 | 1972 |
|---|---|---|---|---|
| Aircraft in operation[a] | 1,220 | 1,867 | 2,390 | 2,347 |
| Average available seats per aircraft | 38 | 66 | 110 | 118 |
| Revenue miles flown (millions) | 370 | 821 | 2,028 | 2,000 |
| Revenue passengers carried (millions) | 17 | 56 | 153 | 172 |
| Fatalities[b] | 109 | 363 | 1 | 185 |
| Passenger fatalities per 100 million miles flown | 1.2 | 0.9 | — | 0.13 |

[a]Excludes helicopters.
[b]Includes crew members.

Source: Statistical Abstract of the United States, 1974.

FIGURE 15.3

Progress in Takeoff Noise Reduction

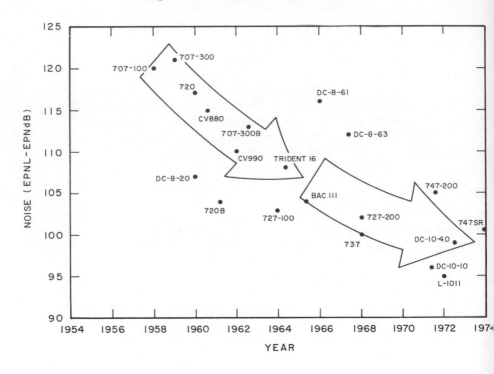

Source: Reference 5.

Noise standards have historically been administered by the FAA, and under pressure of FAR Part 36 [18], the noise characteristics of new aircraft are greatly improved over their predecessors [5], as shown in Figure 15.3. Recently, however, to be consistent with its jurisdiction over other noise sources [19],* the EPA has been given strong jurisdiction over aircraft noise. The EPA may propose noise regulations to the FAA, which is bound to publish them and either adopt them or state reasons why it has not within a fixed period of time. It is safe to assume that a government agency will be involved in aircraft noise regulation whenever hydrogen aircraft might begin service.

*This move has resulted in interagency jurisdictional rivalry that will surely be resolved before a hydrogen-fueled aircraft design begins.

The most common and fairly sophisticated tool of aircraft and airport noise management is the Noise Exposure Forecast (NEF) [18, 19]. The weighted factors that influence NEF ratings are:

Absolute noise level
Noise frequency spectrum
Maximum tone
Noise duration
Aircraft type
Mix of aircraft
Number of operations
Runway utilization
Flight path
Operating procedures
Time of day

A hypothetical NEF 30 contour for a particular mix of aircraft is shown in Figure 15.4. The NEF 30 contour is particularly significant because it corresponds to the boundary within which land use for residential purposes is normally considered unacceptable [20].

Each individual airplane operating at a particular airport has its own noise contour or "footprint." The NEF contours for an airport are the weighted sum of all the contours of the airplanes using the airport, taking into account all the other factors listed above [21]. Since the scale employed in the addition is logarithmic (based on the decibel scale, dB), the noisiest aircraft dominate the composite NEF contour. As Figure 15.3 showed, the oldest aircraft are the noisest and, since these are generally the first to be replaced, introduction of any new aircraft will tend to reduce the area within the NEF 30 contour (unless an increased number of operations offsets the improvement in the individual airplanes) [22].

Air pollutant emissions also come under the regulatory authority of the EPA. The clean burning aspects of the hydrogen-fueled airplane could conceivably lead EPA to increase the stringency of aircraft emissions standards until use of hydrogen became the best alternative to controlling emissions available to the aviation industry. However, recent experience with the relaxation of the timetable for automotive emissions clean-up because of claimed hardship on the part of the automotive industry makes it doubtful that emissions standards will be made that strict.

Several agencies of government play important roles in the aviation system. The CAB, established in 1938, regulates the rates, routes, service entry/abandonment, and mergers for the 43 scheduled and nonscheduled interstate airlines [16]. The CAB also regulates U.S. operations of international airlines. A similar role is played for intrastate airlines by state public utilities commissions.

## FIGURE 15.4

## NEF Contours for a Representative Single–Runway Airport in 1970

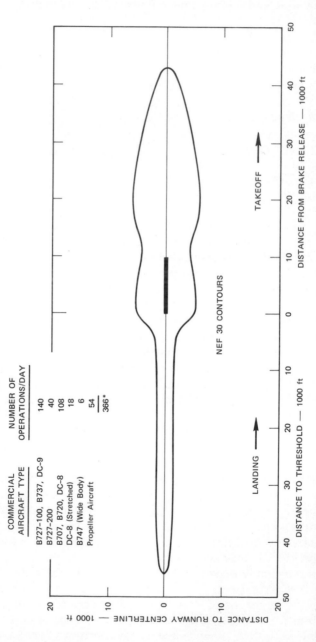

**REPRESENTATIVE LARGE AIRPORT (1970)**

| COMMERCIAL AIRCRAFT TYPE | NUMBER OF OPERATIONS/DAY |
|---|---|
| B727-100, B737, DC-9 | 140 |
| B727-200 | 40 |
| B707, B720, DC-8 | 108 |
| DC-8 (Stretched) | 18 |
| B747 (Wide Body) | 6 |
| Propeller Aircraft | 54 |
| | 366* |

NEF 30 CONTOURS

*Total operations, 86.3 percent of which occur, during daytime (0700–2200 hours) and 13.7 percent of which occur during nighttime (2200–0700 hours).

<u>Source:</u>  Adapted from reference 19.

226

The CAB would become a participant in a change of aircraft fuel indirectly through its rate-setting capacity in the event that a differential in fuel cost affected the revenue and profitability of airlines. The CAB would play a more direct role when the existence (or absence) of hydrogen fuel facilities affected the viability of the route structure of regulated airlines. Since the mission of the CAB is both to protect the consumer interest of the traveling public and to foster a healthy and vigorous air travel system, it would be concerned whether a transition to hydrogen might unduly alter the competitive position of an airline with poor access to hydrogen.

## TRANSITION TO HYDROGEN

To achieve a transition from the present petroleum-based aircraft fuel to hydrogen would be a major undertaking, and the risks for all the stakeholders would be major. Yet, as indicated above, in the aviation sector the number of decision makers involved is far fewer than in the automotive sector, and this will tend to simplify the decision-making process. Moreover, as also indicated by the discussions above, the stakeholders have a long tradition of close cooperation and highly coordinated decision making, whereas in the automotive sector, manufacturers cannot directly coordinate their decisions with their consumers. Consequently, it is probable that if a hydrogen economy were to be realized, aviation would be the sector involving merchant hydrogen most likely to make the first large-scale significant shift.

Deliveries of the next generation of subsonic passenger airplanes may not occur before 1990-2000.* In the introduction of any technology that requires an infrastructure—such as the aviation fuel production and distribution networks—there is a difficult period of coordination reminiscent of the query of whether the chicken or the egg came first. Systems that are familiar today and that require vast infrastructures were introduced slowly and incrementally as demand grew from nearly zero. The transition to hydrogen-fueled airplanes would have to be much more rapid to justify economic participation by most of the major

---

*Because the newer jets use improved materials and construction techniques, unlike the early jet planes, they have no definite lifetime determined by structural fatigue. As a result, the historical periodicity in the turnover of airplane fleets may have been broken. Unless compelling advances in the aviation state of the art occur to motivate replacement, the present generation of wide-body aircraft could continue in service to 1990 or beyond.

parties having an interest. Consequently, a nearly unprecedented coordination of effort would be required [15].

An implementation plan can be envisioned in which the airplane manufacturers, the airlines, the fuel suppliers, and the government coordinate efforts to make aircraft and fuel available at essentially the same time. A planned implementation schedule could probably enable the hydrogen-fueled system to supplant most of the long-haul jet-fueled system within a decade once deliveries began. The scenario might proceed as follows:

1. In 1976 public and private discussions about the transition might begin.
2. In 1980 the decision to produce both an airplane (perhaps selected in a government competition) and fuel would be made.
3. In 1995 the first liquid-hydrogen-fueled aircraft would be delivered to airlines, and fuel would be available at four of the most separated major air travel hubs, probably New York, Washington, D.C., San Francisco, and Los Angeles.
4. Between 1995 and 2000 one new airport could be added to the network every six months within the United States, and facilities for fueling would be installed at major overseas airports, such as London, Amsterdam, Frankfurt, and Rome, thereby making possible the first international flights.

No attempt need be made to drive the transition to completion, for when the transition had been started, there would be strong incentive for all parties involved to convert the trunk lines to hydrogen as quickly as possible to avoid unnecessary costs from duplicated facilities. It is not likely that feeder airlines would be in a position to use hydrogen, but they could await development of a smaller airplane with sufficient hydrogen-carrying capacity to enable loop routes to serve small airports for which provision of liquid hydrogen would be uneconomical.

Insight into the feasibility of a transition to a hydrogen-fueled commercial aviation industry is aided by examination of the size of the required system elements and comparison with the natural building blocks. The top 5 air travel hubs (New York, Los Angeles, Chicago, San Francisco, and Atlanta) account for about 36 percent of the national passenger total, and the top 25 airports account for about 75 percent [1]. As an example, San Francisco International Airport (SFO) is considered in Table 15.3, where it can also be seen that the costs of completely switching the fuel supply and distribution system of even a single airport over to hydrogen are very large. Although most of the costs shown in Table 15.3 would have to be borne by the fuel suppliers, these costs greatly exceed the total expenditures of $1.3 billion in 1971 by federal, state, and local governments on airport expansion or alteration. The large size of these investments will surely result in the exercise of great caution in approaching a decision about hydrogen-fueled aviation.

To add perspective to the investment needs established in Table 15.3, the investment that would be required to produce a synthetic crude oil suitable to produce synthetic hydrocarbon jet fuel (identical to that now used) from coal and oil shale would be in the range of $500 to $700 million 1973 dollars* for a 100,000-barrel-per-day plant [23, 24]. This is based on the 1973 jet fuel usage at SFO, which averaged 40,000 barrels per day [25], and assumes that existing refineries would process the coal or oil shale synthetic crude oil and that existing distribution pipelines could be used [26]. The major reason why the synthetic crude oil approach to providing aviation fuel proves less costly is that much of the existing fuel-processing system would remain useful, while the nuclear/electrolysis hydrogen system would require a fresh start.

The lower investment cost for synthetic jet fuel compared with that for liquid hydrogen naturally raises the question of why the hydrogen option should be entertained at all. The answer, quite obviously, lies in the finite nature of the coal and oil shale resources—ultimately these resources will decline in accessibility and a transition to another source of fuel will be required. However, U.S. shale and coal reserves will last for a century or considerably longer, † depending on the total growth of energy consumption in the United States [27], future accessibility of domestic and foreign crude oil, and the extent to which coal resources are developed for other uses (for example, coal gasification, liquefaction, and electric generation). Given that someday a change will probably be necessary, the basic question facing aviation is the choice of the most propitious time to invest in a change.

The various stakeholders are likely to hold conflicting opinions about the desirability of the transition. The major concerns of the various stakeholders regarding a hydrogen-fueled aviation sector will probably be the following:

---

*Plant construction costs are estimates, since no actual commercial-sized plant has been built or contracted. As Figure 11.1 showed, the plant construction index has escalated far more rapidly than other indexes in recent years. Thus, recent news items cite significantly higher costs in current dollars. However, the same escalations would affect the investments cited in Table 15.3.

†But since many foreign countries do not have such large reserves of nonpetroleum hydrocarbons, U.S. aircraft makers might respond to their needs.

TABLE 15.3

Airport Liquid Hydrogen System
(San Francisco International, 1973 use rates)

| Supply of Demand Component | Average Hydrogen Energy Flow of Building Block (GW)[a] | Number of Building Blocks Required | Unit Capital (millions of dollars) | Total Capital (millions of dollars) |
|---|---|---|---|---|
| Fuel demand | 2.5 | — | — | — |
| Fuel system | | | | |
| Fuel supply options | | | | |
|   Nuclear/electrolysis | 0.53 | 5 | | |
|   Nuclear power plants | 1.0 | 5 | 560.0 | 2,800 |
|   Electrolytic plants | — | — | 34.0 | 170 |
|   Coal gasification | 1.5 | 2 | 140.0[b] | 280 |
| Hydrogen gas pipeline (30-inch diameter) | 2.7 | 1 | | Depends on distance |
| Liquefaction facility | 0.32 | 8 | 22.5 | 180 |
| Totals | | | | |
|   Nuclear option | — | — | — | 3,150 |
|   Coal gasification option, excluding pipeline | — | — | — | 630 |

[a]See Table 12.1.
[b]See Linde's estimate, Table 11.4.

Airlines
    Economics of all facets
    Public acceptance (safety)
    Assurance of fuel supply
Aircraft manufacturers (without government subsidy)
    Added costs of development
    Effect on foreign sales
Airport operators
    Costs of airport alteration
    Safety to people at the airport
    Noise levels
    Air pollution emissions
Fuel producers and distributors
    Relative investment requirements and profitability of hydrogen
       compared to other fuel supply alternatives
    Protection of still valuable investments in the hydrocarbon fuel
       system
    Ability to assure customers of supplies
Federal Aviation Administration
    Safety of the fuel in the aircraft and during fueling
    Noise levels
Civil Aeronautics Board
    Airline financial health
    Cost of air travel
Environmental Protection Agency
    Noise levels
    Air pollution improvements
Federal government in general
    Energy policy effects of hydrogen in aviation
    Balance of trade effects of fuel and aircraft

These concerns would probably lead to the following grouping, according to degree of enthusiasm for hydrogen in aviation:

Most enthusiastic (in descending order)
    Hydrogen producers*
    EPA
    Aircraft manufacturers
    Airlines
Basically neutral
    FAA

---

*But probably only if they are not also producers of conventional jet fuels.

CAB
Airport operators
Least enthusiastic
Fuel producers (conventional fuels)

Before these lines are drawn even as firmly as the above categorization would imply, considerable hardware research and development must be achieved to alleviate some major uncertainties in the practicality (as opposed to feasibility) of hydrogen-fueled aircraft [14], and systems studies must be performed to indicate how the transition would have to be phased to prove successful. Because the military is apparently not supporting any significant research and development in hydrogen-fueled aviation [11], and because aircraft manufacturers cannot afford it in the amounts needed, governmental research and development funding appears to be a necessary prerequisite for reaching the point where the hard, potentially expensive decisions can be made on rational bases.

Whether the aviation component of implementation schedule scenarios depicted in Figures 13.1 through 13.4 comes to pass depends crucially on the new information generated and the perspective gained in the next decade.

## IMPACTS

Although it cannot be foreseen whether a confluence of events will actually result in the introduction of hydrogen-fueled commercial aviation near the end of this century, the rest of this chapter discusses the impacts of the hypothesized transition.

### Economic

Efficient utilization of the large capital investment in airplanes is essential to profitable airline operation. Since planes can generate revenue only when they are flying, the airlines try to minimize maintenance and turnaround times. The key is to have adequate ground facilities to handle schedule peaks. Clearly, the introduction of liquid hydrogen at airports would have to be coordinated with the advent of the hydrogen plane. The large investment in each element of the aviation system would be expensive to bear if one element were forced to remain idle while waiting for the other to be deployed. The transition to hydrogen aircraft would differ from the introduction of new generations in the past: A totally new infrastructure would be required for fueling. The fuel system from production through distribution would

have to be deployed. The lead times for the fuel system would equal
or exceed plane development time, especially for the nuclear-based
option. From an airline viewpoint, the coordination problem might be
overwhelming, since the airline would not be likely to have control
over either the fuel or the aircraft end of the situation. The airlines
would be open to great risks if forced to pay for one part of the system
while waiting for the other parts.

Other factors that affect the airlines' decision to purchase new
aircraft are the used-aircraft market, the depreciated value of existing fleets, and the availability of capital. Because of the short depreciation periods taken by the airlines for tax purposes, the existing
fleets are seldom physically "worn out" by the time enough depreciation allowances have been accumulated to afford replacement [9].*
The used-aircraft market will depend greatly on the worldwide availability and price of liquid hydrocarbon fuels. If the emergence of hydrogen-fueled aircraft were attendant on an absolute, worsening shortage of hydrocarbon fuels in the United States, the domestic market for
used hydrocarbon-fueled aircraft would be poor, but if hydrogen-fueled
aircraft represented a response to a policy of national energy independence while ample supplies of foreign petroleum still remained, the
market for used aircraft would not be affected much. The two aircraft depreciation rates controlled by the Internal Revenue Service
(IRS) and the CAB are strong factors in the ability of airlines to generate internal funds and to show a profit. Naturally, the ability to attract external capital depends largely on the profitability of an airline.
Accelerated depreciation of new hydrogen-fueled aircraft is a potential
governmental policy mechanism by which an airline could generate
profits and attract external capital, both of which could be used for the
acquisition of new aircraft.

The financing of airport facilities is related to the airlines'
financial situation. Currently, state and local authorities issue revenue bonds based partly on landing and use fees collected from the airlines. Therefore, the feasibility of issuing such bonds (as well as
their interest rate) ultimately depends on the financial capability of
the airlines. Moreover, at some airports, notably O'Hare in Chicago,
airlines agree to underwrite residual costs (that is, costs after all
revenues from concessions, parking, and so on have been collected)
[30].

Table 7.1 showed that the initial purchase price for a hydrogen-
fueled airplane is expected to be essentially identical to that of an advanced jet-fueled airplane, but the energy utilization, measured on an

---

*However, for rate-setting purposes, the CAB allows a deprecia-
tion based on the expected time to technological obsolescence [30].

energy per seat-mile basis, is expected to be lower (about 5 to 10 percent) [14]. Thus, an airline could presumably afford to pay a slight premium for hydrogen fuel and still come out even with a jet-fueled plane [14].

Two other key factors of economic importance will be the turnaround time of the aircraft at an airport and the reliability and maintenance schedule. Since the time for refueling apparently is not the pacing factor in aircraft turnaround today, unless hydrogen refueling proved to be significantly slower, this aspect should pose no direct problem. However, it has been suggested, presumably to enhance safety and lessen the delivery costs for liquid hydrogen, that liquid hydrogen airplanes should refuel and undergo between-flight cabin refurbishing at a central "island" removed from the airport terminal. This system is currently in use at Dulles International near Washington, D.C., where passengers are shuttled between the terminal [28] and and the airplanes in special vans [28]. If this procedure proved a necessity rather than an option, the average turnaround time at airports might be increased, thereby adversely affecting the aircraft utilization factor. Whether this system would increase or decrease costs of airport modification is unclear.

Maintenance of hydrogen-fueled aircraft engines should be simpler and less frequent than for jet-fueled engines because the fact that hydrogen mixes so quickly and completely with air and burns so cleanly allows use of a simpler combustion chamber [29]. These characteristics also result in lower maximum local temperatures (about 1,800°K compared with 3,000°K), allow the use of less exotic materials [29], and reduce engine wear. The maintenance of liquid hydrogen fuel storage and delivery systems in the airplane, however, may be more frequent (at least initially) and more complex than for conventional jet fuel systems [14]. The challenge of designing fuel systems with maintenance intervals competitive with conventional jet fuel systems can probably be met.

A key constraining force on airlines embarking on the use of hydrogen would come from the complex system of aircraft routing employed. Seldom is an airplane merely shuttled back and forth indefinitely along a fixed route. Airlines seek to optimize their passenger load factors and equipment utilization to remain profitable. Considerations of travelers' preferences for times of departure and arrival* and complications caused by differing time zones, airport curfews, aircraft maintenance schedules (relative to the location of airline maintenance bases), and rules about the length of work for air crews all necessitate the optimization of the complete complex system. As

---

*This varies by month, day of the week, and time of day.

a result, an individual airplane may journey from Boston to Houston
to Albuquerque to Phoenix to San Diego to Chicago to New York, with
days elapsing before the cycle is begun again [30].

At the beginning of a transition to hydrogen only a few airports
could be equipped to refuel aircraft [15], and this would imply simple
shuttlelike routing of aircraft that would result in a degree of aircraft
underutilization that would affect profitability. Even as the number
of airports dispensing hydrogen increased, the mix of two kinds of aircraft in the fleet and the continual change in the mix and number of
fuel-dispensing locations would tend to impede optimization of equipment utilization. Unless the public showed great enthusiasm and preference for hydrogen-fueled planes and thereby improved load factors, *
it would seem natural for each individual airline to want to hold back
until the pioneering with hydrogen had been accomplished by its competitors.

Aircraft manufacturers clearly have much at stake in the choice
of whether to make a hydrogen-fueled aircraft. Their risk is much increased by the lack of research and development for hydrogen-fueled
military aircraft transferrable to a civilian aviation. Historically,
military aircraft led the way technologically, and this enabled manufacturers to lessen the development costs for civilian aircraft. For
example, much of the design for the Boeing 747 was accomplished in a
military transport design competition, which Boeing lost to Lockheed's
design of the C5-A.

Nevertheless, aircraft manufacturers have considerable experience in liquid hydrogen technology from their work as prime contractors
for space launch vehicle stages.

There is some question whether any single manufacturer would
dare to develop a hydrogen-fueled airplane out of step with its competitors, who might remain with conventional jet fuel because they
anticipate development of synthetic jet fuel derived from coal and oil
shale. There is no good historical analogy, for even when Boeing and
Douglas took the lead in introducing commercial jets, all the competitive manufacturers had military jet experience and the military jet
fuel distribution network was already extensively deployed. There is
room for a difference of opinion about whether being first with a hydrogen airplane design would yield rewards commensurate with the risk.

Besides being concerned with domestic sales, the aircraft manufacturers must be very concerned about international sales prospects.
Historically, foreign airlines have relied heavily on U.S.-made aircraft in both the new and the used markets. Conversely, U.S. aircraft
producers have relied heavily on foreign sales. Table 15.4 shows the

---

*Load factors play a critical role in airline profitability [31].

TABLE 15.4

Role of Commercial and Civilian Aircraft in U.S. Foreign Trade

| Cateogry | Year | | |
|---|---|---|---|
| | 1965 | 1970 | 1972 |
| Aircraft, parts and accessories (value in $ millions) | 1,798 | 3,053 | 3,494 |
| New passenger transports (number of units) | 61 | 149 | 96 |
| New cargo transports (number of units) | 4 | 7 | 17 |
| Used, rebuilt, modified, converted* (number of units) | 406 | 358 | 450 |

*Includes changes from military to nonmilitary type.

Source: Statistical Abstract of the United States, 1974.

extent of this dependence. It appears that if the foreign air carriers also switched to hydrogen-fueled aircraft, the dominance of U.S. manufacturers would not change. But if U.S. carriers were to switch unilaterally to hydrogen-fueled aircraft, the market would be split, so that it could prove difficult for a U.S. manufacturer to make money on either a jet-fueled aircraft or a new hydrogen-fueled design, unless there were great commonality in design. Since many foreign countries are too small to justify use of long-haul hydrogen-fueled airplanes domestically, and since much of their overseas traffic is to and from the United States, they might readily agree to go along with a change to hydrogen for long-haul aircraft.

Sales to foreign airlines might not be necessary for production of a profitable plane, provided that fewer than three companies were competing for the domestic market [9], but the larger the market, the less the risk to the manufacturer. The use of hydrogen airplanes on international routes would clearly depend on development of fuel facilities at major overseas airports.

The question of what would induce foreign governments to move to a hydrogen fuel base for air transportation* is a subject for concern to the aircraft industry. Just as in the United States, foreign interests

---

*Such a move might create market opportunities for the U.S. nuclear power industry.

could be expected to respond to a shortage of hydrocarbon fuels or to increasing prices of these fuels, but they would not necessarily be responsive to U.S. policy of energy independence. Instead, each nation could be expected to view its own energy policy on its own merits.

Several kinds of U.S. governmental policy intervention could apply pressure on foreign carriers serving the United States; for example, hydrocarbon fuel could be refused to foreign carriers, or the government could tax hydrocarbon fuels prohibitively. Alternatively, the use of hydrocarbon fuels could be banned on the basis of pollutant emission rates, much as the United States has banned supersonic flights of SSTs over populated areas because of the sonic boom. Of course, such measures might evoke retaliation, which could cripple international air transportation and cause serious diplomatic problems.

Maintenance of U.S. dominance of the air transport market is a factor in the decision to produce a new airplane that might stimulate governmental intervention with subsidies or incentives. There are two underlying considerations: the balance of trade and national defense. In combination, these two considerations may result in an expressed U.S. policy to subsidize airframe manufacturers or airlines as an incentive to produce a new generation of aircraft and to maintain production capabilities and vitality in an industry important to national defense. Alternatively, by appealing to foreign governmental concern about their own balance of payments, the prospect of a joint hydrogen-fueled aircraft production venture might prove a very potent inducement.

Employment opportunities generated by a transition to hydrogen aircraft fall into two main categories: with airplane manufacturers and with fuel producers and distributors. Presumably, it will continue to be government policy to maintain a healthy aerospace industry in the interests of national defense. Manufacture of aircraft and parts provided employment for 514,000 people (of whom 281,000 were production workers) in 1973 [32]. These jobs should be relatively unchanged either in skill needs or in geographic distribution by a transition to a hydrogen-fueled aviation system. An estimated 20,000 people are engaged in the refueling of commercial aircraft. These jobs would have altered skill requirements if a transition to hydrogen occurred. The largest effect would be the transfer of jobs from the petroleum industry to the hydrogen fuel industry and the basic energy resource (such as nuclear power) industry.

## Environmental

The most obvious environmental consequence of a transition to hydrogen-fueled aircraft is the reduction of emissions of air pollu-

tants [14, 29]. The estimates shown in Table 15.5 were developed
in Lockheed's comparative evaluation of future subsonic conventional
jet-fueled and hydrogen-fueled passenger transports. Hydrogen data
for carbon monoxide, unburned hydrocarbons, and smoke are firm
since the complete absence of carbon compounds in liquid hydrogen
precludes the possibility of their formation. The data on nitrogen
oxide emissions, however, are not firm because there have been no
adequate experimental measurements of this parameter. Yet, theoretical considerations indicate that nitrogen oxide emissions will be
quite low in an engine specially designed for hydrogen [29]. This
reduction stems both from the faster and more thorough fuel mixing
and burning, which eliminates the localized hot spots most responsible for generating $NO_x$, and also from reduction of residence time in
the combustor [29].

   The doubling of water vapor emitted from a hydrogen engine
compared with that emitted from conventional jet-fueled airplanes is
a relatively minor problem because in the troposphere (where subsonic jets fly) water vapor is not considered a pollutant in the usual
sense. However, along heavily traveled air lanes, exhaust contrails
(condensed water vapor) can add to cloud cover. This has been noted

TABLE 15.5

Air Pollutant Emissions Comparison: $LH_2$ Versus Jet-A Subsonic
Passenger Aircraft

| Emission Product | Engine Condition | Estimated Emission Level (g/kG fuel except as shown) | |
| --- | --- | --- | --- |
| | | Jet-A | $LH_2$ |
| CO | Idle | 30 | 0 |
| Unburned HC | Idle | 4 | 0 |
| Smoke | Takeoff | 15[a] | 0 |
| $NO_x$ | Takeoff | 12 | $\leq 12$[b] |
| $H_2O$ | Cruise | 41.9 lb/nmi | 82.4 lb/nmi |
| Odors | Ground operations | Objectionable | None |

[a]SAE 1179 smoke number
[b]Adjusted for the difference between the gravimetric energy densities of Jet-A and hydrogen (a factor of 2.8 in favor of hydrogen).

Source: Reference 14.

informally by a researcher at the National Center for Atmospheric
Research (NCAR) in Boulder, Colorado, which is under a major east-
west air lane [33].

The pollutant emissions from subsonic aircraft while in transit
make a relatively insignificant impact on overall air quality. However, air pollution from airplanes can be substantial on the ground
and in the air near airports [42]. When air pollution became an issue,
the public was quick to note emissions of black smoke from aircraft
on takeoff, and, as a result, actions to eliminate most of this smoke
have been completed. The emissions of vaporized fuel and unburned
hydrocarbons, however, have not been completely eliminated, and
these are responsible for the fuel odors at airports. Although the
public at large cannot readily sense the emissions of carbon monoxide
and nitrogen oxides, air pollution control measures have also been
taken in recent years to lessen the on-the-ground emissions of these
pollutants. This reduction has been accomplished by altering the taxiing and engine run-up patterns and operations of airplanes on the
ground [34]. The interaction between air pollution control and noise
control is important, because cutting back on the number of engines
used in taxiing and the length of pre-takeoff engine run-up have reduced
both irritants. Thus, the positive air pollution consequences of hydrogen-fueled aircraft would be, by far, most pronounced at and near airports.

Just as in automotive use of hydrogen, the pollutant reduction at
the point of end-use is gained by creating pollution at the point of hydrogen generation. While the trade-off may be favorable from a national
point of view, the results would not be viewed identically in all the regions affected. Moreover, the kind of environmental pollutants are
not necessarily measurable in commensurate terms, thereby making
the trade-off more a matter for political resolution. The clearest example of incommensurate pollutants is the reduction of conventional
air pollutants gained at the expense of increasing the emissions and
risks of nuclear power for hydrogen production.

Noise is the other environmental parameter of most significance
in aviation. In the Lockheed comparative evaluation of future jet-
fueled and hydrogen-fueled subsonic aircraft [14], it was concluded
that both kinds of future airplane would exhibit improvements over
any commercial transport now flying (the quietest is the Lockheed
L-1011, as shown in Figure 15.3). Table 15.6 shows the estimates
Lockheed obtained compared with the present applicable Federal Aviation Regulations concerning noise (FAR 36) [14]. As Table 15.6 shows,
both kinds of airplane are expected to be significantly quieter than the
L-1011 (a change of about 10 dB is equivalent to a factor of two in
perceived noise) [22].

TABLE 15.6

Noise Comparison:  $LH_2$  Versus Jet-A Subsonic Passenger Aircraft

| Aircraft | Noise Levels in EPNdB ( ) = FAR 36 Limits | | | Area of 90 EPNdB Contour (sq mi) |
| | Takeoff | Sideline | Approach | |
| --- | --- | --- | --- | --- |
| 3,000 nmi | | | | |
| $LH_2$ | 88.1 (103.8) | 86.4 (106.3) | 97.9 (106.3) | 3.8 |
| Jet-A | 92.7 (105.1) | 86.4 (106.9) | 96.6 (106.9) | 4.1 |
| 5,500 nmi | | | | |
| $LH_2$ | 89.2 (104.9) | 87.2 (106.8) | 98.4 (106.8) | 4.3 |
| Jet-A | 94.2 (107) | 87.8 (107.6) | 96.7 (107.6) | 4.7 |
| Lockheed L-1011 (certification tests) | 96.0 (105.6) | 95.0 (107) | 102.8 (107) | 6.6 |

Source:  Reference 14.

Because the NEF contours for any given airport are especially sensitive to the noisiest aircraft operations, and the noisiest planes are the old first-generation jets that are still in service, any new transport (using either fuel) will tend to replace the oldest jets and thereby effect a major change in airport noise levels [22]. As Table 15.6 shows, the hydrogen-fueled plane would contribute more to a noise reduction in the takeoff zone but would contribute less in the landing approach zone; effects in the sideline zone are nearly identical for the two kinds of airplanes.

The underlying reason why a hydrogen plane would be quieter in the takeoff zone but slightly noisier in the approach zone can be traced to the relative ratios of fuel weight to aircraft empty weight (see Table 7.1) [35]. Because of the lightness of the hydrogen fuel, the gross takeoff weight of the hydrogen-fueled plane is far less than for the jet-fueled plane. This means that the engine need not be as powerful, and less energy need be expended to get the plane airborne. Since noise output is related to the energy expended by the engine, the noise is less. However, during descent for landing, when all but emergency reserves of fuel have been exhausted, the gross landing weight of the hydrogen airplane is much nearer to its gross takeoff weight than is true of a jet-fueled plane. As a result, during descent, engines of the hydrogen-fueled airplane would have to be run at a proportionately higher power level than those of a jet-fueled airplane. This accounts for the relative noise emissions of the two airplanes. It is important to note, however, that noise differences arise from differences in performance and are not intrinsic to the fuel.

Thus, while a hydrogen-fueled airplane would be quieter than present airplanes, so would a future jet-fueled airplane, and the advantages of the hydrogen plane would be felt most in the takeoff zone. For some airports this could be a major advantage, but for other airports, where the takeoff is regularly over uninhabited areas,* there would be very little apparent advantage of a hydrogen-fueled plane with respect to noise.

### Resource Utilization

As noted earlier in this chapter, the major alternative to hydrogen in aviation in the future is the production of synthetic jet fuel from coal or oil shale. Since, as noted in Chapter 11, coal is the cheapest intermediate-term source of hydrogen (in the near term, natural gas and oil remain cheaper sources), a comparison of coal resource utilization between these two alternatives is important. Figure 15.5, which compares the relative resource utilization effectiveness of jet fuel (kerosene)† and hydrogen from coal, shows that either option uses the coal resource with approximately equal effectiveness. (Although the hydrogen option is shown as a little less effective, this distinction is probably not meaningful within the accuracy of the estimate.)

It is important that to date the DoD shows little enthusiasm for hydrogen-fueled aircraft [11]. DoD's dominance of aircraft research and development efforts make its future actions especially relevant. If DoD were to encourage the production of synthetic jet fuel from coal or oil shale, even perhaps by direct subsidy, the synthetic liquid fuels industry would receive substantial impetus toward development. This might delay serious interest in hydrogen for commercial aviation indefinitely. DoD already has begun to examine the suitability of synthetic fuels derived from coal for meeting its needs [36].

---

*At Los Angeles International, for example, the very regular wind conditions allow takeoff over the ocean nearly all the time.

†It is assumed that the refinery product is not restricted to kerosene alone but the usual product slate (about 6 percent kerosene) is made. Nevertheless, through an exchange of products, the production of jet fuel can be viewed as equivalent to conversion of the synthetic crude to the single jet fuel product.

FIGURE 15.5

Resource Utilization Comparison for Synthetic Jet Fuel (Kerosene) from Coal and Liquid Hydrogen from Coal

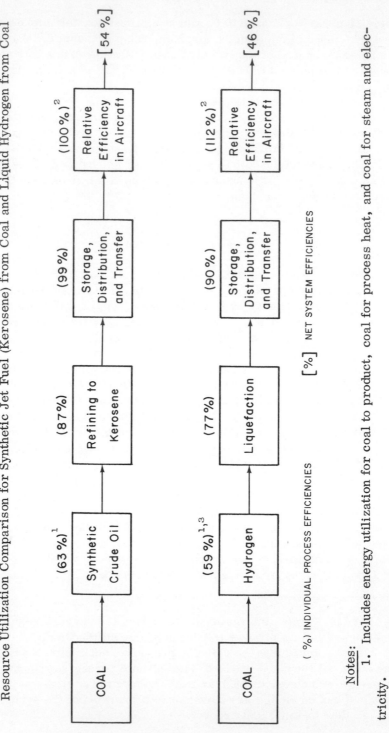

( % ) INDIVIDUAL PROCESS EFFICIENCIES    [ % ]  NET SYSTEM EFFICIENCIES

Notes:
1. Includes energy utilization for coal to product, coal for process heat, and coal for steam and electricity.
2. Reference 14.
3. SRI internal data.

242

Although it is sometimes thought that a major advantage of the conversion of aviation to hydrogen would be the release of petroleum for other uses, such as automobiles, Figure 15.5 shows that this option is really no more effective in this respect than the direct production of synthetic automotive fuels from coal.*

## Social

About one of four Americans over 18 will fly this year. People who fly tend to be between 20 and 50 years of age and to have more income and education than those who do not fly. About half the trips are taken for business and convention purposes and the other half for personal reasons of visiting and sightseeing. When people have flown, they tend to fly again and to perceive airplanes and airports as good, economically desirable, and timesaving. In 1972, the average air passenger took five or six trips. People who fly on business take more trips per year than people who fly for personal reasons. It seems quite possible that the demand for air travel may decrease owing to increased energy prices. Additionally, the public may have a change in life-style and in values that could further decrease this demand.

The energy situation that hit the American home and automobile for the first time in 1973-74 is likely to dampen the growth of both the supply and the demand for air travel well before the turn of the century, when the first hydrogen-fueled passenger aircraft could be operating. For both philosophical and economic reasons, people may seek to substitute less energy-consumptive modes of transportation for air travel. Also, by the year 2000, telecommunication of information and entertainment may compete with travel services through such devices as the videotelephone [37] and cable television.

The energy and oil shortages could affect the U.S. demand for air travel in less direct ways, such as through significant shifts in employment patterns, decreased employment, lower average real income, and higher real prices of commodities that will ripple outward in the economy as a result of increased energy prices. With the decline of discretionary income, proportionately fewer passengers would be able to afford personal air travel, and life-styles may shift to greater emphasis on necessities and the lower-cost luxuries. A sentiment against air travel could even arise from an energy-conscious public, who might view energy-intensive air travel as siphoning off petroleum products

---

*The efficiencies for producing jet or automotive fuel from coal are essentially the same.

from their cars.  However, if the expected improved efficiency of hydrogen airplanes were realized in practice, then the public might wel-
come the development—especially if it were perceived as releasing
petroleum for use in the automobile.*  Thus, for many reasons, it is
plausible to expect air traffic in the United States to grow at rates
considerably less than current rates and less than those that have
been frequently projected for the next 20 to 30 years by stakeholders
in the commercial aviation sector.

Noise has become recognized as a source of detrimental health
effects [21, 38].  In fact, noise levels have been used as indicators
of the quality of life in some studies.  The positive health benefits derived from decreased aviation noise (however slight) will have the
most effect on persons working in or living near airports, since they
are most exposed to noise.  The magnitude of this impact would increase as the transition to hydrogen aircraft became more pronounced.


## Social Effects of Safety

One of the major social effects of hydrogen-fueled aircraft will
be the safety of individuals and groups who are involved with the aircraft either actively or passively.  Early in a transition to hydrogen
aircraft it will not be possible to give proof of safety as determined
by airline experience.  Groups with special need for concern include
passengers, air crews, ground crews, airport operators, and airport
neighbors.  The last three groups can be expected to view the matter
with some sophistication once adequate actual data about hydrogen
safety in aviation has been accumulated.  Therefore, the discrepancy
between the perceived and the actual safety of hydrogen can be expected
to be least for these three groups.

Nevertheless, air crews and ground crews will have to be trained
and informed about the relative hazards of hydrogen, and about the
ramifications for their own and passenger safety in the event of a malfunction.  These groups will have to be convinced of the safety of handling hydrogen-fueled aircraft, since they will be directly encountering
the potential hazards.  If these groups are not convinced of the safety
of hydrogen, they may attempt to block implementation through asso-
ciations and labor unions.  In addition, emergency service crews (such
as airport fire departments) will require special training to cover the
new hazards associated with hydrogen.  During the early stages of
transition, special precautions would probably be required until a large
amount of experience was accrued.

---

*Although Figure 15.5 shows that this perception could be ill
founded.

Passengers, however, who do not have the opportunity or need to be trained in the safe handling of hydrogen, would probably view the safety of the hydrogen aircraft less realistically than the crew operating it. Yet, being more educated and younger than the average citizen, the airline passenger might know something about the new fuel, maybe even enough to convince himself that the level of risk was acceptable. Probably, however, the passenger will not squarely face whatever facts are available at the time on his chances for arriving safety at his destination. Instead, today's average airline passenger accepts risks to safety with a sense of fatality. It has been found that few passengers want to see the cockpit, for example, because it reminds them of the fallibility of the human beings piloting the airplane [39].

Just as the passenger prefers not to know directly the risks of flying, or the precautions taken to reduce those risks, neither does the airline make explicit reference to its safety records when advertising. Instead of using data of the type given in Table 15.7 in their advertising [40], airlines prefer to use "prestige" advertising as a way of building up the image of their airline as sound, safe, and reliable.

Airlines may not be able to avoid public attention to the safety issue once hydrogen-fueled aircraft have been introduced. Most people are likely to rely on their friends and acquaintances rather than on experts for verification of the real safety hazards. The spread of erroneous information could easily be started by news media linkage of hydrogen-fueled aircraft with the Hindenburg and with other disasters, especially since the movie in the "disaster" genre, Hindenburg, will have become an "old" film suitable for frequent showing on television by the 1990s, when hydrogen-fueled airplanes might be introduced.

Environmental action groups, while lauding the clean air and noise reduction aspects of the hydrogen airplane on the one hand, may

TABLE 15.7

Death Rates from Transportation Accidents

| Travel Mode | Fatalities per 100 Million Passenger Miles |
|---|---|
| Airlines | 0.10 |
| Motor buses | 0.08 |
| Railroad | 0.56 |
| Automobiles | 1.9 |

Source: Reference 40.

take exception to the coal or nuclear power source of hydrogen on the other hand and choose to deprecate hydrogen's safety as an argument against airport expansions. This could strike a responsive chord in the general populace, because with the aircraft flying overhead the potential would exist for anyone to be exposed to the hazards. It seems likely that introduction of a hydrogen-fueled airplane would have to be accompanied with a widespread education program to alleviate public apprehension. Such a program must be planned very carefully to ensure against the possibility of its backfiring through the spread of incomplete information or information that does not take pains to address any misleading material that may be presented in the disaster movie.

Residents near airports would have special concern over the safety of the aircraft. Because of their location, they are confronted daily with the realities of low flying and noise and air pollution from aircraft. Naturally, they tend to have more concern than the public in general over the possibility of a crash occurring within their neighborhood. The introduction of hydrogen-fueled aircraft would probably stimulate complaints from airport neighbors. Partly because it represented a change, they might look on it as an opportunity to reiterate their dissatisfaction with life near the airport.

This fear on the part of the residents near airports has ramifications for their perception of the noise from a hydrogen plane. Although, as shown in Table 15.6, the actual noise exposure from hydrogen aircraft would be less, research has shown that tolerance to noise levels is related to individual feelings about airports. Sociological interview techniques were applied in a study of the noise problems in 22 U.S. communities near military airfields [38, 41]. Individuals who not only deemed the air base to be important and considerate, but also expressed little fear about aircraft crashes, tolerated four times the daily noise exposure before complaining as individuals who both were fearful and had negative feelings about the air base and its importance. A study of civilian aircraft noise revealed that, of eight factors contributing to annoyance by noise, the most important was "fear of aircraft crashing in the neighborhood" [38, 42]. Even though with hydrogen-fueled airplanes the actual noise level would decrease, nearby residents might still react to their apprehension by complaining about noise. Thus, it is also likely that the number of noise complaints would decrease as experience proved the hydrogen aircraft safe.

## Technology

The development of hydrogen-related technologies suited for use in aviation would have a spin-off to other possible hydrogen-using sectors. The large fuel demands of aviation would provide substantial stimulus to advancements in the state of the art in hydrogen production, distribution in liquid form, and cryogenic storage. Thus, although in principle the aviation sector could be the only sector to use hydrogen, the effects of its use would probably spread outward and enhance the likelihood that other sectors would also ultimately embrace hydrogen.

## GOVERNMENTAL ROLE

A hydrogen-fueled aviation system appears to make a great deal of sense, although many technical, institutional, social, environmental, and economic issues need to be understood more clearly before private or public decisions to support a hydrogen-based commercial aviation industry would be justified. Indeed, some of the nontechnical issues may prove to be critical bottlenecks. In general, however, the prospects for switching aviation to hydrogen seem much better than those for switching automobiles, especially since the decision-making process would be more focused and the implementation could be more easily programmed.

The magnitude of the development work needed before a transition to hydrogen-fueled aviation could begin appears to be beyond the capabilities for internal sponsorship by the aerospace industry. Moreover, the risks of such sponsorship are great because the questions that must be faced in anticipation of deployment require an unprecedented degree of coordination among the various stakeholders—with a very uneven distribution of economic and political power among them.

Recognizing the difference between a decision to explore the option and a decision to deploy the option, the federal government's leadership and sponsorship appear to be needed to address some of the research and development questions and to explore the feasibility and possible strategies of implementation. Such strategies must clearly recognize the perceptions and needs of the various stakeholders. Only the government appears capable of initiating the dialogue in a serious fashion, since the other stakeholders are likely to remain snagged on the "chicken or egg" dilemma because of the economic risks involved.

The federal government's interest in the possibility of hydrogen-fueled aviation would be found along several lines:

Air pollution
Noise abatement
Safety
Cost of travel (rate setting)
Energy resource utilization

TABLE 15.8

Summary of Impacts of Hydrogen–Fueled Aircraft

| Impact Class | Stakeholders | Nature of Effect | Magnitude of Impact (units) |
|---|---|---|---|
| **Economic** | | | |
| Investment required for production and distribution of hydrogen fuel | Cryogenic industry<br>Airports<br>Oil companies<br>Utilities<br>Capital market | Large demand for capital to supplant system<br>Vying for roles in system | Major ($) |
| Investment required to produce a new class of aircraft | Airframe manufacturers<br>Airlines<br>Subsystem manufacturers | Large demand for capital to develop system | Major ($) |
| Cost of aircraft fuel, maintenance, and repair | Cryogenic industry<br>Airlines<br>Air passengers | Personal and business travel possibly restricted by higher costs | Moderate ($, people) |
| **Institutional** | | | |
| Regulatory | Aircraft industry<br>Government regulatory bodies<br>Component producers<br>Airline pilots | Expanded regulatory authority over design criteria for performance and safety of operation | Major (power) |
| Ownership of fuel system | Airports<br>Public energy utilities<br>Oil industry<br>Pipeline companies<br>Cryogenic industry | Competition for control of hydrogen fuel markets | Moderate (power) |
| Land use | Developers<br>Conservationists<br>Nuclear plant builders<br>Municipalities<br>Airport neighbors | Competition for use of land in vicinity of airport for new fuel supply system | Major (power, $) |

## Social

| Factor | Stakeholders | Description | Impact |
|---|---|---|---|
| Air travel demand | Passengers<br>Airlines<br>Aircraft industry<br>Competing transportation or communication | Effort by energy-conscious travelers to find lower energy-consumption options to reduce cost | Moderate (people, $) |
| Employment | Skilled workers<br>Construction workers<br>Cryogenic engineers<br>Nuclear power industry<br>Coal gasification industry | Employment availability in aircraft production, airport construction, power, coal, and hydrogen industries | Moderate (people) |
| Airport noise | Airport neighbors<br>Airports | Transitory rise in complaints that are fear-induced, followed by long-term decrease | Moderate (people) |
| Safety | Air and ground crews<br>Airports<br>Airlines<br>Passengers<br>Airport neighbors<br>General public | Public attention to aircraft safety<br>Opposition to airport growth | Major (people, $) |

## Environmental

| Factor | Stakeholders | Description | Impact |
|---|---|---|---|
| Air pollution | Airport neighbors<br>Public at large<br>EPA | Greatly reduced air pollution at point of energy consumption | Moderate (people, $) |
| Noise | Airport neighbors<br>Airlines<br>Airport operators | Reduced noise, especially in the takeoff zone | Minor (people) |
| Trade-off | Public at large<br>Regional interests<br>EPA | Shifting of pollution from point of use to point of production | Moderate (power, people) |

## Governmental

| Factor | Stakeholders | Description | Impact |
|---|---|---|---|
| National defense | Public at large | Maintenance of strong aerospace industry | Major (power, $) |
| National defense/balance of trade | Public at large | Shift from foreign energy supplies to domestic sources | Major (power, $) |

Airworthiness
Balance of trade

The absence of previous significant government expenditures for hydrogen-fueled military aircraft means that the usual indirect—but very real—subsidy of commercial aircraft is lacking. Accordingly, an economic risk-mitigation or financing arrangement similar to that initiated for the supersonic transport might be warranted should studies demonstrate that deployment of hydrogen-fueled commercial aircraft is desirable.


## SUMMARY

There are many complex economic, institutional, social, and environmental consequences of a transition to hydrogen-fueled commercial aviation. Table 15.8 summarizes some of these.


## REFERENCES

1. "The Long Range Needs of Aviation," a report of the Aviation Advisory Commission, Washington, D.C., January 1973.

2. U.S., Department of Transportation, Federal Aviation Administration, FAA Statistical Handbook of Aviation, Calendar Year 1973.

3. "Cyclical Aircraft Introduction May Slow," Aviation Week and Space Technology, October 28, 1974, p. 42.

4. "Next Generation Transports Will Emphasize Fuel Savings," Aviation Week and Space Technology, October 28, 1974, pp. 48-51.

5. J. E. Steiner, "The Technology and Economics of Commercial Airplane Design—Part I," presented to the Swedish Academy of Aeronautics and Astronautics, Stockholm, November 8, 1972.

6. "Stretched DC-10 Aimed at Fuel Saving," Aviation Week and Space Technology, October 28, 1974, pp. 32-33.

7. D. V. Maddalon, "Rating Aircraft on Energy," Astronautics and Aeronautics, December 1974, pp. 26-40.

8. "Fuel Outlook Dictating Technical Transport Research," Aviation Week and Space Technology, October 28, 1974, pp. 52-63.

9. Grayden Paul, Lockheed California Company, Burbank, Calif., personal communication, 1974.

10. "Lockheed Spruces up Its Financial Image," Business Week, May 26, 1975, pp. 29-30.

11. Robert Ziem, Office of the Director of Defense Research and Engineering, Department of Defense, Washington, D.C., personal communication, 1974.

12. G. Hill, "The Clamor Against Noise," San Franciso Chronicle, September 15, 1972. Also New York Times News Service.

13. H. B. Safeer, "Aircraft Noise Reduction—Alternatives Versus Cost," Sound and Vibration, October 1973, pp. 22-27.

14. G. D. Brewer et al., "Study of the Application of Hydrogen Fuel to Long Range Subsonic Transport Aircraft," National Aeronautics and Space Administration, Langley Research Center, Hampton, Va., January 1975.

15. C. R. Dyer, M. Z. Sincoff, P. D. Cribbins, eds., "The Energy Dilemma and Its Impact on Air Transportation," National Aeronautics and Space Administration-American Society for Engineering Education, Langley Research Center, Hampton, Va., 1973.

16. "Transportation—Freedom from Regulation?" Business Week, May 12, 1975, pp. 74-80.

17. Edward Versaw and Harold Bradley, Lockheed California Company, Burbank, Calif., personal communication, 1974.

18. J. K. Power, "Aircraft Noise Standards and Regulations," Federal Aviation Administration, April 1971.

19. "Transportation Noise and Noise from Equipment Powered by Internal Combustion Engines," Office of Noise Abatement and Control, U.S. Environmental Protection Agency, December 31, 1971.

20. R. D. Beland, P. P. Mann et al., "Aircraft Noise Impact: Planning Guidelines for Local Agencies," Department of Housing and Urban Development, November 1972.

21. C. R. Bragdon, Noise Pollution (Philadelphia: University of Pennsylvania Press, 1971).

22. James Young, Stanford Research Institute, Menlo Park, Calif., personal communication, 1974.

23. F. H. Kant et al., "Feasibility Study of Alternative Fuels for Automotive Transportation," vol. I: Executive Summary; vol. II: Technical Section; vol. III: Appendices, Exxon Research and Engineering Company, Linden, N.J., for the U.S. Environmental Protection Agency, June 1974.

24. J. Pangborn and J. Gillis, "Alternative Fuels for Automotive Transportation—A Feasibility Study," vol. I: Executive Summary; vol. II: Technical Section; vol. III: Appendices, Institute of Gas Technology, Chicago, Ill., for the U.S. Environmental Protection Agency, June 1974.

25. William V. Paizis, Airports Commission, City and County of San Francisco, personal communication, 1975.

26. E. Dickson et al., "Impacts of Synthetic Liquid Fuel Development for the Automotive Market," Stanford Research Institute, Menlo Park, Calif., for the Energy Research and Development Administration, Washington, D.C. (in preparation).

27. "U.S. Energy Prospects: An Engineering Viewpoint," Task Force on Energy of the National Academy of Engineering, Washington, D.C., 1974.

28. J. E. Johnson, "The Economics of Liquid Hydrogen Supply for Air Transportation," in Advances in Cryogenic Engineering, vol. 19, ed. K. D. Timmerhaus (New York: Plenum, 1974), pp. 12-22.

29. A. Ferri and A. Agnone, "The Jet Engine Design That Can Drastically Reduce Oxides of Nitrogen," paper no. 74-160, presented at the American Institute of Aeronautics, Twelfth Aerospace Sciences Meeting, Washington, D.C., January 30-February 1, 1974.

30. James Gorham, Stanford Research Institute, Menlo Park, Calif., personal communication, 1974.

31. L. J. Carter, "Airlines: Half-Empty Planes Keep Profits Low, Waste," Science 184, no. 4139 (May 24, 1974): 881-84.

32. Statistical Abstract of the United States, 1974 (Washington, D.C.: Bureau of the Census, U.S. Department of Commerce, July 1974).

33. James Lodge, National Center for Atmospheric Research, Boulder, Colo., personal communication, 1970.

34. P. Martin and V. Salmon, "South Florida Airport Site Selection Study: Document Aircraft and High Speed Ground Transport Noise and Pollution Impact Factors," Stanford Research Institute Project MSH 1181-957, Menlo Park, Calif., July 1971.

35. G. Daniel Brewer, Lockheed California Company, Burbank, Calif., personal communication, 1974.

36. R. L. Goen et al., "Synthetic Petroleum for Department of Defense Use," Stanford Research Institute for the Defense Advanced Research Projects Agency and Air Force Aero Propulsion Laboratory, Wright Patterson Air Force Base, Ohio, Technical Report AFAPG-74-115, November 1974.

37. E. Dickson and R. Bowers, The Video Telephone: Impact of a New Era in Telecommunications (New York: Praeger, 1974).

38. K. D. Kryter, The Effects of Noise on Man (New York: Academic Press, 1970).

39. E. Dicter, Handbook of Consumer Motivation (New York: McGraw-Hill, 1964).

40. "Air Transport Facts and Figures 1973," Air Transport Association of America, Washington, D.C., 1973.

41. P. N. Borsky, "Community Reactions to Air Force Noise. I. Basic Concepts and Preliminary Methodology. II. Data on Community Studies and Their Interpretation," Report TR 60-689 (II), National Opinion Research Center, University of Chicago, Chicago, Ill., 1961.

42. TRACOR, "Community Reaction to Airport Noise," final report, TRACOR Company, Austin, Texas, 1970.

# 16

## INTRODUCTION

Three classes of energy utilities can be expected to play an important role in a hydrogen economy: gas, electric, and combined electric and gas.

A utility is an unusual form of enterprise because it is a monopoly countenanced by the law and the public. Moreover, it is a "natural" monopoly in the sense that it is impractical, both logistically and economically, for two utilities to offer the same services in any given area. To prevent the abuse of monopoly power, regulatory commissions have been established to control the utilities [1]. Most large utilities are private, but there are some large government-owned utilities, such as the Tennessee Valley Authority and the Bonneville Power Administration, and numerous municipally owned utilities.

Regulatory commissions at the state and federal level have power to control the rates utilities charge customers [1]. State regulatory bodies are generally called public utilities commissions. At the federal level, the Federal Power Commission (FPC) is concerned with the interstate movement of gas and electric power. To change rates, a utility must petition the relevant regulatory commission and justify the proposed rate change in public hearings [1]. Such hearings are often slow and long delayed in starting [2, 3]. The regulatory commissions also have jurisdiction over facilities siting.

The rates utilities charge customers are determined according to a "rate-base" formula that allows a utility a certain maximum percentage return on its capital investment [1]; the range is normally 7 to 9 percent. In practice, the utility definition of "capital investment" is somewhat different than in other industries. In particular, some items, such as interest paid during construction of a facility, that other industries might treat as an expense are considered capital investment by utilities because plant cost is not figured into the rate base until it becomes operational.

Utilities, being natural monopolies, have certain understood obligations to the public [1]. First, they must remain financially healthy because their services are essential to public welfare and no alternatives to their service are immediately available. Consequently, the regulatory commissions are concerned not only with direct consumer welfare but also with indirect consumer welfare as manifested in the health of utility industry. Second, in return for protected status as a natural monopoly, utilities have historically been obligated to serve all applicants, to provide nondiscriminatory service, and to refrain from interfering with the nature of the consumer's use of services (unless, of course, the use posed a threat to the integrity of the distribution system). For example, electric utilities have been obligated to provide electric power whenever and in whatever quantities customers decided (and without warning) [2]. However, because of the large number of customers served, these uncertainties in demand tend to average out, thereby making instantaneous response manageable. Nevertheless, utilities are required to maintain a certain degree of ready reserve to meet statistically rare peak demands [2, 4].

Since utilities have long been regulated natural monopolies with a fairly steady rate of return on investment, their securities (stocks and bonds) are often regarded as safe, "blue chip" investments [5].* Moreover, this same steady, sure business has enabled utilities to attract capital especially effectively because the risks to investors have been low. The nature of the heavy equipment involved and their ability to attract capital have led utilities to become the most capital-intensive large industry in the United States. Energy utilities accounted for about 11 percent of all U.S. business capital investment in the decade 1961-71 [5].

A reduction in utility profitability and ability to attract capital has taken place in recent years, and this has become a matter of grave public policy concern [2, 5]. The rising costs of utility construction (at a rate that exceeds the general rate of inflation—see Figure 11.1), and the soaring costs of borrowed money (interest) have left utilities unable to attract all the capital they require to add facilities to meet increasing consumer demands for energy [2, 5]. Moreover, the diminishing supply of natural gas and the rising costs of oil and coal have eroded the profitability of utilities as their costs rose much faster than their service rates could follow. This has been ex-

---

*Recent declines in profitability and skipped dividends have tarnished this image, however.

acerbated because the consumer outcry over utility rate increases has greatly extended the approval process, thereby increasing the lag between rising costs and recovery through rate increases [3].

There have been other changes from the historical norm for utilities as well. The diminished availability of natural gas, for example, has also tempered the requirement to serve all potential customers; in some service areas, gas utilities have been allowed to refuse to connect newly constructed buildings into the service network.

Hydrogen's future role in the utilities affects several of these new conditions in important ways:

1. A transition to hydrogen would enable gas utilities to deliver indefinitely a gaseous energy carrier to consumers [6].

2. Hydrogen could be a cost-effective way for electric utilities to store energy, thereby avoiding the need for some new and expensive generating capacity [4].

3. Hydrogen facilities will themselves be capital-intensive, thereby continuing the problem of capital attraction.

The impacts of a hydrogen economy on the three classes of utilities would be diverse and far-reaching.

## GAS UTILITIES

### Structure and Stakeholders

The natural gas industry is composed of a number of stakeholders with the utility providing only the final link between production and the consumer. The major institutional stakeholders are the following:

1. Gas companies that produce gas as their main business
2. Oil companies that produce gas as a by-product
3. Gas transmission pipeline companies
4. Gas utilities
5. Governmental regulatory agencies

The relationship of these stakeholders is shown in Figure 16.1.

Oil companies generally regard gas production as a sideline and sell most of the by-product gas produced (some is used in their own operations) rather than become involved in gas delivery to consumers. Gas companies regard natural gas production as their main business and explore and develop potential natural gas fields; any liquid by-products are considered a sideline. Historically, pipeline companies

FIGURE 16.1

Stakeholders in the Gas Industry

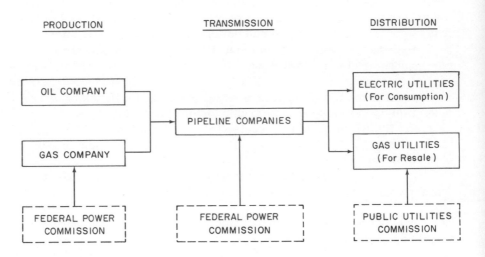

have been operated as the transportation mode to convey gas from the region of production to the utilities for distribution to final consumers. Pipeline companies are often owned jointly by several producing companies or utilities.  Gas is sold to the pipeline company as it enters the pipeline and then sold to the utility as it exits the pipeline.  Thus, natural gas pipelines differ from other forms of public transportation carriers (such as railroads) because they own the commodity conveyed. As domestic natural gas production declines, trunk pipelines are becoming underutilized.  But, since there is little wear on these pipelines, they will last for many more years; the pipeline companies, accordingly, are concerned with keeping the pipelines operating at full capacity to maintain the profitability of their investments.  As a result, much of the recent interest in coal gasification to produce synthetic methane (SNG) has been shown by pipeline companies.  For example, El Paso Natural Gas Company, a pipeline company, is committed to build a large coal gasification plant in the Four Corners region of New Mexico [7].

Gas utilities serve many customers all around the nation. There are several classes of service, but the most important distinctions are between "firm" and "interruptible" contracts [9, 10].  Customers with firm contracts have the highest priority for gas supplies, while deliveries to interruptible customers can be curtailed during peak demand periods if supplies fall below demand [9, 10]. In return for accepting the risk of interruption on short warning, these customers pay lower rates than firm customers—often only a tenth as much [11]. Residential and most commercial use is in the firm category, while much industrial use is in the interruptible category [10]. When gas supplies were abundant, the difference in risk between these classes of service was more theoretical than actual because interruptions were rare. In practice, the distinction usually proved to be merely a means to provide large industrial users with lower-priced gas. Recently, however, the risks of curtailment have been real, and widespread interruptions have occurred [9, 10]. Now many industrial users are reconsidering the trade-offs involved in their choice of fuel.

## Consumers

The geographic distribution of gas consumers is shown in Table 16.1 by class of service. Comparison of the gas pipelines shown in Figure 6.1 and Table 16.1 leads to several observations: First, total gas consumption is highest in the regions nearest its production (especially in the Gulf Coast region) and in the highly populated, industrialized Appalachian/Mid-Atlantic and Great Lakes regions. It is noteworthy, however, that in the latter regions residential use greatly exceeds use of the industrial type (the sum of "firm" industrial and nonutility interruptible). Second, industrial use in the Gulf Coast gas-producing states is exceptionally high, showing that when natural gas is easily available (and relatively cheap compared with other fuels) it is the choice industrial fuel. Third, the total use of gas in electric power generation is about 17 percent of the total consumption. Fourth, industrial use ("firm" industrial plus nonutility interruptible) accounts for about 43 percent of all gas consumption, but only 0.4 percent of all customers. Fifth, if the disproportionate industrial use in the Gulf Coast region were reduced to a level similar to industrial use in other regions, and if all natural gas used to generate electric power were eliminated, there would be an effective extension of natural gas reserves by about 30 percent.* This is largely the reason behind the federal energy policy to displace natural gas from electric utility

---

*Data from Table 16.1:

$$\frac{\text{Gas saved}}{\text{Total gas used}} = \frac{2,700 + (2,100 + 1,500)}{21,000} = 30 \text{ percent}$$

boilers [10] and from industries that could use other fuels. Sixth, residential and commercial use currently accounts for about 34 percent of all use but would account for about 50 percent if the conditions of the fifth observation above pertained.

Since the natural gas users generally most able to convert to alternative fuels (such as coal) are industries and electric utilities, and those least able to convert are residences and small commercial establishments, the remaining natural gas supplies will probably be allocated by federal policy according to the following curtailment priorities [10, 11]:*

    Highest
        Residential
        Commercial
    Intermediate
        Ammonia producers
        Petrochemical industry
    Least
        Industrial heat
        Electric utility heat


### Electric Power Generation

On the face of it, the likely places to expect hydrogen first as a natural gas replacement would seem to be in electric generation and industrial uses. On closer consideration, however, the prospects for burning hydrogen to generate electricity (except in load leveling discussed later in this chapter) are very slim, because if the hydrogen were obtained by thermochemical or electrolytic processes in the first place this would prove a circuitous and inefficient way to generate electricity for delivery to final demand. Only if the hydrogen were obtained from coal gasification would its use to make electricity make much sense. Moreover, because the coal could be burned directly by the utility after a suitable (even though costly) plant conversion, the choice of hydrogen would, in reality, represent a decision to transport the energy content of coal by gas pipeline rather than as solid coal in a train. Alternatively, it could represent a deliberate approach to air pollution control. † Thus, it must be concluded that

---

*See Appendix to this chapter for details of the current curtailment priorities.

†Utilities are showing interest in shipping coal by rail or slurry pipeline to the utility for on-site gasification to a low-heating-value

TABLE 16.1

U.S. Gas Consumption by Region and Class of Service, 1973

(billions [$10^9$] of cubic feet at 1,000 Btu per cubic foot at 14.73 psia)

| | Firm | | | Utility Power Generation | | Interruptible[a] | Other | Total[b] |
|---|---|---|---|---|---|---|---|---|
| | Residential | Commercial | Industrial | Firm | Interruptible | | | |
| U.S. total (excluding field use) | 5,000 | 2,300 | 6,200 | 2,100 | 1,500 | 2,300 | 1,400 | 21,000 |
| New England[c] | 140 | 55 | 36 | — | 6.2 | 18 | 11 | 270 |
| Appalachian/Mid-Atlantic[c] | 1,500 | 610 | 100 | 17 | 120 | 280 | 250 | 3,800 |
| Southeast[c] | 270 | 160 | 300 | 130 | 120 | 520 | 130 | 1,600 |
| Great Lakes[c] | 1,100 | 55 | 90 | 62 | 35 | 270 | 130 | 3,000 |
| Northern Plains[c] | 260 | 130 | 160 | 4.7 | 180 | 200 | 60 | 990 |
| Mid-continent[c] | 320 | 130 | 160 | 280 | 230 | 250 | 160 | 1,500 |
| Gulf Coast[c] | 440 | 203 | 3,300 | 1,500 | 320 | 310 | 500 | 6,500 |
| Rocky Mountain[c] | 180 | 110 | 40 | 17 | 46 | 180 | 18 | 580 |
| Pacific Southwest[c] | 730 | 260 | 210 | 120 | 500 | 600 | 120 | 2,500 |
| Pacific Northwest[c] | 69 | 44 | 83 | 2.5 | 0.2 | 150 | 6.7 | 350 |
| Pacific[c] | — | — | — | — | — | — | — | 74 |

[a]Other than utility power generation.
[b]Totals may not add because of data rounding.
[c]

| New England | Appalachian/Mid-Atlantic | Southeast | Great Lakes | Northern Plains | Mid-continent |
|---|---|---|---|---|---|
| Connecticut | Delaware | Alabama | Illinois | Iowa | Kansas |
| Maine | District of Columbia | Florida | Indiana | Minnesota | Missouri |
| Massachusetts | Kentucky | Georgia | Michigan | Nebraska | Oklahoma |
| New Hampshire | Maryland | North Carolina | Wisconsin | North Dakota | |
| Rhode Island | New Jersey | South Carolina | | South Dakota | Pacific Southwest |
| Vermont | New York | Tennessee | Gulf Coast | | Arizona |
| | Ohio | | Arkansas | Rocky Mountain | California |
| Pacific Northwest | Pennsylvania | Pacific | Louisiana | Colorado | Nevada |
| Idaho | Virginia | Alaska | Mississippi | Montana | New Mexico |
| Oregon | West Virginia | Hawaii | Texas | Utah | |
| Washington | | | | Wyoming | |

Source: Reference 8.

the application of gaseous hydrogen is more likely in industry than in electrical generation.

## Industrial Use

Use of natural gas in industry in 1968 is depicted in Figure 16.2, which shows that about 63 percent of all natural gas consumed by industry is used to make process steam and only about 30 percent is used in direct heat applications [12]. Much of the process steam is low-temperature steam.

Since virtually any alternative energy form (electricity, coal, oil, solar, and so on) could generate process steam, to capture this part of the market, hydrogen would have to prove cheaper (on a unit energy basis), more convenient, or cleaner. However, as discussed in Chapter 11, since hydrogen is a secondary or derived energy form, it is unlikely to be cheaper than alternatives for a very long time. If hydrogen were available today, in many instances it would prove more convenient and less costly to convert equipment from natural gas combustion to hydrogen combustion rather than to switch to other fuels. But delays inherent in making such a large transition will ensure that hydrogen cannot be made generally available to industry for many years. Consequently, conversions are likely to be made to other energy forms long before hydrogen becomes a realistic choice for industry. In the future, hydrogen's most potent competitor would be electricity, because it is a clean, reliable, and versatile energy source.

Realistically, therefore, hydrogen's future as an energy form for steam generation in industry seems restricted to applications requiring very high-temperature steam. As discussed in Chapter 7, the combustion product of hydrogen is steam (water vapor). Moreover, if hydrogen is burned in an atmosphere of pure oxygen, the resultant steam can have a very high temperature and be exceedingly free of contaminants.

Doubtless, there are many applications where the burning of a gaseous fuel is the most straightforward manner to apply direct heat. Hydrogen's chances of capturing this market would appear to be excellent unless other heating approaches (such as induction or microwave)

---

mixture of carbon monoxide, hydrogen, and (noncombustible) nitrogen. The gas is then consumed with improved thermal efficiency in an adjacent electrical generation plant that employs an advanced combined-cycle system. Besides the gain in thermal efficiency, the ability to isolate and control most of the air pollutant emissions of coal at the gasification plant instead of at the power plant is attractive.

FIGURE 16.2

Industrial Use of Natural Gas, 1968

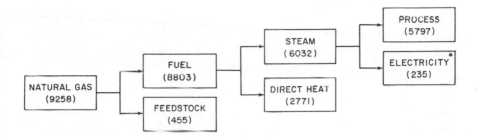

*Does not include natural gas used to generate electricity in util-
ities.

Note: Numbers in parentheses are quantities in trillions ($10^{12}$)
of Btu.

Source: Reference 12.

had already become the accepted norm in the interim between the near-
term reduction in the industrial use of natural gas and the long-term
potential use of hydrogen.

Barriers and Impacts

There are important logistical barriers to the delivery of hydrogen to industrial markets as long as the residential and commercial
markets remain on natural gas. First, two separate (but often parallel)
delivery systems would have to be operated. This implies added
capital and operating expense compared to a single system. Moreover,
with two separate systems, economies of scale could not be realized as
fully as with a single, larger system. This adds substantially to the
already strong arguments in favor of the gas utilities embracing SNG
rather than hydrogen. Second, with the industrial and residential-
commercial uses disengaged in the gas system, the shape of the daily,
weekly, and annual load profiles would change. Presumably, with the
implied loss of diversity in end-uses, the load profiles of both the hydrogen and natural gas portions of the utility would exhibit more accen-

tuated peaks and valleys than is now the case. This, in turn, would imply the need for additional capital expenditures to ensure the delivery of peak-load gas supplies.

These logistical and financial barriers have important implications for the delivery of hydrogen to industrial users but methane to residential and commercial users. First, there would be a strong incentive to avoid the introduction of hydrogen at all; but if it were introduced, the awkwardness and penalties of a dual system would provide an equally strong incentive to complete the transition and also shift residential and commercial users to hydrogen.*

This total conversion would be a step that utilities could not take lightly, for in addition to requiring a change on the part of every consumer (his appliances, probably his gas meter, leak-tightening his pipes, and so on) [6], there would be a large expense incurred in upgrading and testing the large and dispersed distribution network that serves residential areas. In 1973, there were about 40 million residential customers and over 600,000 miles of gas distribution mains in the United States [13]. The labor needed to convert the high-capacity system that serves concentrated industrial users would be small compared to the labor needed to convert the dispersed low-capacity system that serves residential areas.

Figure 6.1 and Table 16.1 together suggest that as methane becomes less available, its distribution would tend to be retrenched to those areas closest to production or where consumption is already the largest. Moreover, since intrastate gas is not regulated by the FPC and is now sold at much higher rates than interstate gas,† there is strong incentive for producers and pipeline companies to retain and sell the gas in the state of origin [14]. Once established, this pattern of gas buying and selling will tend to resist quick change. Thus, it can be concluded that it is in the northern portion of the western United States where gas utilities might first have to contemplate seriously a change to hydrogen.**

The availability of hydrogen or natural gas on a local or regional basis would affect industry decisions about the location of new plants,

---

*The only exceptions would appear to be those where large, new heavy industry parks were established with a large, new gaseous hydrogen supply separate from the other system from the very beginning.

†This situation is a subject of much public policy debate, and the regulated price of interstate gas is expected to be increased [14].

**There are important external variables, however. First, the delivery of natural gas from Alaska [15] could either come via a trans-Canadian pipeline for delivery into this area, or in the form of LNG brought by tankers to western ports. Second, there is a good possibility that LNG will be imported to the West Coast from Siberia.

or the relocation of marginal plants. If hydrogen were to be perceived as the fuel of the future by industrialists and were competitively priced, its availability would probably serve as a regional attractant. But if, as seems most likely, the price of hydrogen were not competitive with alternative energy forms such as coal or electricity, the conversion to hydrogen in a region would probably chase industry to areas where energy was cheaper. There are historical analogues. For example, after World War II, labor-intensive textile manufacturing migrated from New England and the Mid-Atlantic states to the southeast because labor was cheaper there. In the future, energy-intensive industries can be expected to contemplate seriously similar moves based on regional energy prices. For companies with marginal operations swept up involuntarily in a conversion, and for whom there were no intrinsic benefits in hydrogen, the expense of conversion might mean ruin.

Besides the larger, institutional, considerations of a transition to hydrogen, there would be impacts on the personal level. About 40 million households (about 60 percent of all those in the United States) are supplied with natural gas [13]. At the residential level, most gas meters and all gas burners would require conversion (see Chapter 7) to operate on hydrogen [6]. Even if performed on a neighborhood-by-neighborhood basis, this process implies that a very large number of people would be inconvenienced. On this basis alone, it seems certain that many individual citizens or citizen lobbies would oppose a transition to hydrogen. Yet, with careful planning this impact could probably be managed quite well. There have been recent experiences of similar large-scale conversions (for example, England's conversion from coal gas to North Sea natural gas over the last few years) that while inconvenient, proved more a topic for conversation than a topic of serious complaint.

During the period of transition when some localities were served by hydrogen at the residential level while others were served with natural gas, the mobility of citizens seeking to move from one city to another would be somewhat impaired because their gas-burning appliances would require reconversion whenever they changed kinds of gas service territory. This limitation is mainly related to stoves, and to a much lesser extent, gas clothes dryers. However, stoves are frequently left behind with the house (often because they are built-in) or apartment (because the stove is normally provided with the dwelling). Consequently, this possible restraint on mobility should not prove a very serious handicap.

For safety, hydrogen used in residences would require the addition of an odorant and a colorant to make leaking gas detectable by smell and flames detectable by sight (see Chapter 8) [6]. During the period of transition, when individuals were least experienced with using hydrogen, safety hazards arising from carelessness would probably be greatest.  Thus, it is during this period of transition that maximum attention would have to be given to public information about hydrogen.

In summary, although at first sight it would seem that a conversion of gas utilities to hydrogen might be a natural first step toward a hydrogen economy, the extremely large numbers of residential customers, the expected priority of allocations for natural gas [10], and the expected barriers to industrial conversion all suggest that the gas utilities will strongly prefer synthetic methane (SNG) to conversion to hydrogen.  As will be seen in the next section, the prospects for hydrogen use (captively) in electric utilities seems a much more attractive and realistic prospect.

## ELECTRIC UTILITIES

### Structure and Stakeholders

The structure of the electric power industry is very different from that of the gas industry, and electric utility problems and opportunities are also very different.  The major stakeholders in the electric power industry are fuel suppliers, equipment suppliers, electric utilities, and regulatory agencies.  Until nuclear fission energy began to be used to generate electric power, utilities either were fueled by coal, oil, and natural gas, or utilized hydropower.  Generation equipment is supplied mainly by several large companies—especially General Electric and Westinghouse.  Because the large equipment suppliers also carried most of the burden of research and development for the electric utility industry, it is not surprising that these same companies are the dominant suppliers of nuclear power plants.

It has become increasingly common for electric utilities to become interconnected in a large "grid" as a means to lessen the problems of meeting variations in demand and to sell spare power to one another whenever it is available.  Consequently, electric utilities have become increasingly interdependent.  Along with the benefits of interconnection of systems have come some disadvantages, especially the need for increasingly sophisticated system control since a "fault" or overload in one utility can propagate a problem to other utilities in less than a second. *  Thus, system stability is a major problem of the increasingly complex grid [4].

_____

*Such a problem was the cause of the 1965 blackout in the northeastern United States.

Electric utilities today face the following major problems [14]:

1. Obtaining capital to finance new facilities
2. Siting of new nuclear power plants
3. Air pollution control of fossil-fueled power plants
4. Obtaining approval for new electric power transmission corridors
5. Smoothing out cyclical variations in demand (load leveling)

The most likely role for hydrogen within the electric utility industry is in load-leveling systems, and this use would, in turn, affect the capital and power plant siting problems. Such application of hydrogen will be discussed later in this chapter.

## Energy Transmission

A less likely, but nevertheless frequently mentioned, role for hydrogen revolves around the problems of power plant siting, transmission corridors, and air pollution control. As indicated in Chapter 7 and illustrated in Figure 7.5, electric utilities could conceivably attack all three problems at once. To avoid the problem of cleaning the air pollutants out of smoke from fossil-fueled electric generation plants, a utility can install nuclear power plants. But it has become increasingly difficult to locate nuclear power plants near demand centers and, as a result, the electricity must be transmitted ever greater distances. Because the technology and expense of underground transmission is unfavorable, such transmission is nearly always accomplished by overhead high-voltage wires strung between large steel towers. Opposition rooted in questions of land use and aesthetics, however, has made it increasingly difficult and time-consuming for utilities to gain new overhead transmission corridors or to expand the capacity of those already in use [16].

Consequently, it is often suggested that instead of transporting the energy in electrical form it could be transported in chemical form (gaseous hydrogen carried in pipelines) [6, 17-20]. As shown in Figure 7.5, the hydrogen could be generated either directly from the nuclear heat by closed thermochemical cycles or by electrolysis of water. Once at the demand center, electricity could be regenerated either by fuel cells [21, 22] or by burning the hydrogen cleanly in turbines. The main advantage of this approach is that the land use and visual impacts of underground hydrogen pipeline transmission are much less than for visually obtrusive overhead transmission towers and cables. Thus, it could prove easier to obtain transmission corridors for hydrogen pipelines than for electric transmission lines [16]. Moreover, flexibility in power plant siting might be gained.

FIGURE 16.3

Cost of Hydrogen Generation and Transmission as a Function of
Distance

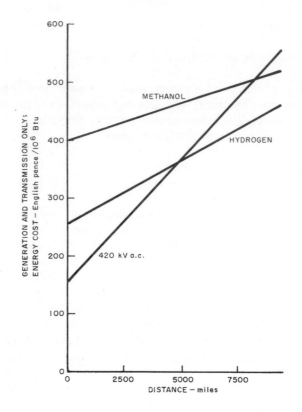

Source: Adapted from reference 24.

There is disagreement in the hydrogen economy literature about
the distance at which energy transmission in the form of hydrogen is
cheaper than overhead electrical transmission [23-25]. Two competing claims are shown in Figures 16.3 and 16.4. In any event, it is
apparent that the distance in question is large—400 miles or more.
Thus, this plan has economic merit mainly for moving power very long
distances. However, as noted in Chapter 7, and illustrated by the
building blocks in Table 12.1, a single nuclear/electrolysis unit would
generate only about one-third of the hydrogen flow needed to operate
the smallest trunk pipeline considered economically practical (24-inch
diameter). When the lowered net efficiency of the total system—be-

FIGURE 16.4

Relative Costs of Transmitting Hydrogen and Electricity

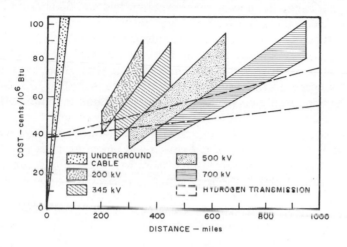

Source: Adapted from reference 25.

cause of the extra conversion losses—is also taken into consideration, it becomes apparent that this plan has little chance of being put into practice for single nuclear power plants. However, because the building-block match is much better, it does seem more attractive for large multiplant "nuclear parks" located in very remote areas.

Load Leveling

By far the most attractive use of hydrogen in utilities is in the form of load leveling [4, 26-28]. As noted in Chapter 7 and illustrated in Figure 7.5, this approach to load leveling involves the use of off-peak electric power to produce hydrogen electrolytically. The hydrogen is then stored and used to regenerate electricity in fuel cells or turbines when peak demands occur. By using off-peak electric power from base-loaded nuclear power, peak-load power might be produced more cheaply than is possible with purchased turbine fuel.* Obviously,

---

*It is important to note that the apparent low cost of this hydrogen can be merely an artifact of internal accounting methods, because the utility can charge itself the incremental cost of the off-peak nuclear

the value ascribed to the hydrogen must at least equal the cost of making and storing it [29].

Although Figure 7.5 showed the possibility of using liquid hydrogen for storage, the use of iron-titanium hydride beds is now widely believed to be the most feasible approach to hydrogen storage in utilities [30]. The iron-titanium hydride is especially well suited to use in utility energy storage because it can be charged and discharged rapidly at pressures and temperatures near that of the hydrogen produced by commercially available electrolyzers [30]. Although the same hydride would be too heavy for general use in automobiles, its weight would be no impediment in this stationary application.

A key parameter in any load-leveling energy storage concept is the round-trip efficiency of energy in and out of storage [4]. In this respect, hydrogen's major competition will most likely come from alakali metal-sulfur batteries, which are expected to be successful in a few years [31]. These batteries are expected to have a high round-trip efficiency, and, as a result, their capital costs can be higher than can a hydrogen storage system and still break even with peak power generated with purchased turbine fuel [4]. Thus, it is by no means certain that hydrogen will be the system ultimately favored by utilities for energy storage.

The use of hydrogen system energy storage would have some significant direct impacts on the electric utilities but would affect customers only indirectly. In particular, the utilities would experience:

1. Increased opportunity for utilities with no access to suitable pumped hydroelectric storage sites to begin load leveling by means of energy storage

2. Increased ability to class more generating equipment as base load and thereby facilitate increased use of nuclear power [4]

3. Improved utilization of generating facilities, thereby improving the cost effectiveness of these large investments [4]

4. Increased system reliability by use of modular hydrogen storage systems [4]

5. Decreased need for "spinning reserve" as normally defined, thereby improving the economics of utility operation [4]

---

power used to make hydrogen. As long as the hydrogen made is reused within the utility, and indeed substitutes for a more expensive fuel for generating off-peak power, the utility saves the consumer money by operating more efficiently. However, should the utility choose to sell the hydrogen for use by other parties, then the cost of electricity used to calculate the value of the hydrogen would have to bear the full average cost of power generation rather than just incremental costs. In that event, the hydrogen would cost more than alternative fuels.

These last two factors are important to utilities but have seldom been mentioned as advantages of hydrogen storage systems (they also appear to be advantages of battery storage systems). A basic requirement imposed on electric utilities is that the probability of a failure be low. To achieve this, utilities have had to build a certain degree of redundancy into their systems and to keep some equipment in a state of ready reserve so that it can produce power within a few seconds of a breakdown of other equipment. This need to keep some generators rotating and synchronized is termed "spinning reserve" [4].

As an example of the extent of spinning reserve, New York State utilities are required to hold the probability of load loss to about 2.5 hours per year [4]. This has been calculated to be equivalent to a 20 percent coincident reserve margin for the New York pool and requires an 18 percent margin for each company based on its own independent peak load [4]. About 5 percent of the margin is classed as spinning reserve that can assume the load in seconds. Gas turbines, which require about five minutes to reach full power, and pumped hydroelectric storage facilities, which require two minutes, cannot be classed as spinning reserve when in a "cold" condition [4].

A hydrogen-fueled fuel cell, however, would be able to reach full power in a few seconds [4].* Consequently, a hydrogen-based load-leveling system that also employed fuel cells would be able to cut spinning reserve costs [4], thereby improving the economics of utility operation and, hence, reducing the cost of power delivered to the consumer. It is also important in this regard that the charge-discharge portions of the hydrogen cycle need not be treated symmetrically. In particular, the storage function could be achieved at a slow rate and even sporadically, while the discharge could take place very rapidly should conditions require it.

Since one goal of utility energy storage systems is to reduce costs, achievement of this goal would adversely affect the industries supplying utilities—especially the fuel suppliers and the electric equipment manufacturers. However, the current and expected continued shortage of energy suggests that fuel suppliers would not be adversely affected,† because latent demand would absorb any slack created by the reduced utility purchases. The traditional equipment suppliers, especially the manufacturers of nuclear power plants, would tend to benefit,

---

*If an iron-titanium metal hydride storage system were unable to yield sufficient hydrogen in just a few seconds, a small store of liquid hydrogen could be held in reserve to cover the start-up.

†Indeed, it appears that anything that relatively painlessly reduces the total growth in fuel demand will affect fuel suppliers beneficially.

because raising the amount of power in the base-load category would create more demand for nuclear power plants. Moreover, a new market for hydrogen storage and conversion equipment would arise, and these same suppliers would probably enter it.

## Load–Leveling Resource Limitations

If iron–titanium metal hydride storage systems were to become popular in electric utilities, there would be a major effect on the titanium industry. To store enough hydrogen to meet 2 percent of the total U.S. delivered electric power by means of hydrogen load-leveling devices would require a total inventory of about $1.6 \times 10^{11}$ lbs of titanium. In 1973, total world production of ilmenite ($FeTiO_3$), the major titanium ore, was about $7 \times 10^6$ lbs, of which 20 percent was from the United States [13]. Since about 31 percent of the weight of ilmenite is titanium, this translates to a 1973 world production of about $2 \times 10^6$ lbs of titanium. Thus, the inventory implied for widespread use of iron–titanium peak storage units is about 100,000 times as large as the present world production capacity for the metal.

Titanium is the ninth most abundant element and is widely distributed. The U.S. Geological Survey estimated titanium reserves as shown in Table 16.2 [32]. Nevertheless, a total need for $1.6 \times 10^{11}$ lbs is more than half the total estimated reserves of $2.8 \times 10^{11}$ lbs [32], and demand of this magnitude would necessarily spur titanium production* to the degree experienced in the past by other critical metal industries (such as copper and aluminum).

## Production of Hydrogen for Sale

Some have begun to mention the possibility of electric utilities manufacturing hydrogen with off-peak power, not for use in energy storage, but for sale [6, 19, 33, 34]. Like production for load leveling, this could lead to a redefined "base load" that would facilitate the introduction of more nuclear power plants (best operated at steady output). Whenever the available base-load power exceeded demand,

---

*It is interesting to note that the most abundant titantium-bearing mineral, ilmenite, already contains iron and titanium in the proportions desired. It is conceivable that hydrogen itself would be used as a reducing agent to remove the oxygen from the compound, thereby possibly greatly simplifying preparation of the iron–titanium host-metal substrate.

TABLE 16.2

Estimated World Reserves of Titanium, 1970

| Country | Titanium $(10^9$ lb) |
|---------|-------------------|
| United States | 50 |
| Canada | 50 |
| USSR | 50 |
| Other countries | 130 |
| Total | 280 |

Source: Reference 32.

the energy would simply be diverted to hydrogen production. Although this concept sounds attractive, simple calculations performed by the Public Service Electric & Gas (PSE&G) utility in New Jersey shows that the quantity of hydrogen produced in this fashion (with the present definition of base load) would only amount to one-fifth the hydrogen output obtainable from a single 1,000MW$_e$ nuclear plant dedicated to hydrogen production [28]. The total energy PSE&G delivered in natural gas form was about 50 times this quantity [35]. Thus, although there has been much discussion in the literature about the possibility of producing hydrogen cheaply by means of low-cost off-peak electric power, the quantities involved would not be large enough to sustain much use of hydrogen. Examination of Table 12.1, however, shows that such generation with off-peak power might be suitable for a few classes of dedicated users, such as a steel mill or an ammonia plant, if sufficiently large storage facilities were available.

Besides involving only small quantities of hydrogen, the concept of producing hydrogen with "cheap" (on the order of 2 mills per kWh) off-peak electric power appears to be based on an allocation of electric power costs that would have doubtful acceptability to regulating agencies. There are two ways of stating the cost of any given unit of electricity produced: the incremental (or marginal) cost and the average cost. Incremental cost treats all the fixed costs of the power plant as if they were already met and as if the only pertinent charges were the fuel and operating costs; naturally, this results in a low unit cost of power [29]. Average cost allocates the fixed charges of all power plant capital investment, taxes, and so on evenly over all power produced and therefore results in a higher figure [29]. At present, most utilities are constrained to charge all customers a rate that does not reflect the time of day that the power is used* and thus does not reflect whether the power consumed is generated by peak-load or base-load equipment [35]. Thus, this rate represents the utility's average cost of producing electric power by all means at its disposal; and under present circumstances utilities could not charge a lower rate for off-peak power for hydrogen production without regulatory approval.† Utility spokesmen have begun to make that point forcefully [4, 29, 36].

A few electric utilities themselves have begun to discuss the possibility of installing nuclear capacity that considerably exceeds their base-load needs with the intention of using the spare capacity to produce hydrogen for sale [28, 34]. There is a hidden presumption that the utilities would somehow charge themselves less for the electric power to electrolyze hydrogen than they would charge other customers, so that hydrogen would be produced at a price competitive with more traditional energy forms. However, it is questionable whether public utilities commissions would or, under public pressure, could allow this form of discriminatory pricing of electric power. Certainly, the net effect of allowing it would be that all customers of the utility involved would be subsidizing the cost of producing hydrogen. In effect, this subsidy would be passed on to the purchaser of the hydrogen. Considering the increased public attention being given to utility rates and public utility commission responses, this issue of cross-subsidy would certainly have to be openly resolved before a utility could risk undertaking a venture that provided a hidden subsidy to hydrogen production.

However, the question of the future price of off-peak power is by no means completely answered. Indeed, there is considerable discussion in the utility industry, by regulators, and in federal energy policy circles about the abandonment of the present rate structure in favor of rates that charge the user more nearly the actual cost of the power he consumes [35]. By this approach, a user would be charged more for power during peak demand periods than during slack demand periods. The goal is to stimulate energy conservation and to use the economic-institutional means of rate structures to achieve both a degree of load leveling of utility demand and abatement of demand

---

*Generally, the rates can vary according to quantity of energy consumed but not according to when it is consumed.

†In any event, the concept itself involves something of a paradox: once hydrogen production became routine, it would have to be viewed as just another part of the base load and thus would no longer be a candidate for special rate treatment.

growth [35]. The proposed rate structures would especially induce those industrial users most able, and those to whom the cost of electric power was the most significant cost of production, to shift their electric loads to slack parts of the day or week. Since the effect of this structure would be to reduce off-peak rates and to raise peak rates compared to the present norm, an institutionally approved mechanism might still arise to make possible hydrogen production with cheaper off-peak power.

Although there are difficult technical problems (such as metering) and political-economic issues involved in instituting "time-of-day" pricing, some European countries have made it a practice. The United States probably will also have made significant moves in this direction before a hydrogen economy makes much headway. It should be noted, moreover, that time-of-day pricing would tend to stimulate the use of electric-powered automobiles that could be recharged at night. In return, this would tend to level the utility load profile by filling in the valleys rather than knocking off the peaks. Thus, the electric car, if satisfactorily developed, not only would give competition to a clean hydrogen-fueled car, but would also pose an indirect threat to the generation of hydrogen with off-peak power.


## COMBINED ELECTRIC AND GAS UTILITIES

In many parts of the United States, gas and electric utilities are combined in one company. The hydrogen economy concept would be attractive to these combined utilities because hydrogen could be used to load-level both the gas and electric systems [27]. As noted in Chapter 7, because of air-conditioning loads, electric utilities generally experience their peak annual demands in the summer, while the gas utilities generally experience their peaks in the winter because of space heating.

Gas utilities must address the problem of where to obtain sufficient hydrogen to allow a conversion in their service areas. Since gas utilities are mainly only involved in passing along to the consumer a commodity in the same form as found in nature, in contrast to electric utilities, they have little experience in the complex technologies of energy conversion. However, in a combined utility, the gas portion of the business could turn to the electric portion for expertise, for example, in nuclear power plant operation. Thus, as far as technical expertise is concerned, it would be easier for combined utilities to move into the use of nuclear power for hydrogen production. As the building blocks in Table 12.1 show, a single nuclear/electrolysis unit could provide enough hydrogen to sustain the residential demands of about 330,000 people.

A combined utility using nuclear power would gain flexibility and a degree of inherent system redundancy that could lead to a reduction in its need to maintain costly reserve systems. This would make possible the delivery of lower-cost energy to consumers. For example, by dedicating some nuclear plants to electric production and some to hydrogen production, but enabling some to swing back and forth between production of the two forms of energy as the demand fluctuates, the utility would be able to use the same nuclear power plants and hydrogen storage systems both to load-level and to provide a substitute for spinning reserve (or the gaseous equivalent) for both systems.

When an electric utility made hydrogen to sell in competition with the gas utility serving the same area, the gas utility could be expected to resist this competition and erosion of its status as a natural monopoly by complaining to the regulatory commissions. A combined utility could sidestep this kind of sticky institutional problem.

The combined utilities that did engage in hydrogen business, and the regulatory bodies that watched over them, would have to resolve the question of possible cross-subsidization of services* to the satisfaction of the public and those who intervene in the public interest. However, since telephone companies and public regulatory utilities commissions have long grappled with this kind of problem [1], no doubt there is a ready paradigm for solution.

Since combined utilities have both technical and institutional advantages compared with separate utilities for participating in a hydrogen economy, it seems likely that the utilities that would be the first to deliver gaseous hydrogen are the combined utilities. Indeed, if the concept of a hydrogen economy gains credence and momentum, this advantage may stimulate mergers between separate gas and electric utilities.

NUCLEAR POWER—A KEY ISSUE

The use of nuclear power to generate hydrogen, or the use of hydrogen as a means of temporary storage of energy produced by nuclear power, is a central underlying theme of this book. Indeed, nuclear power, solar energy, and coal are the three key energy sources for producing hydrogen. However, the preponderance of current thinking on the hydrogen economy concept emphasizes nuclear power to drive a combination electric-hydrogen economy in the long term. The

---

*Where one service is made to pay more than its fair share of costs to allow another service to be priced lower than bearing its fair share of costs would indicate.

quantities of nuclear power implied for hydrogen production alone are enormous. Table 16.3 gives a rough estimate of the number of 1-GW (electric) nuclear reactor building blocks that would be needed to supply the 1973 U.S. energy demands in a few selected sectors (assuming that electrolysis is the approach to hydrogen generation). The total electric generating capacity required for a complete conversion of the aviation, automotive, and residential gas sectors (at 1973 energy demand levels) is 650 plants of 1-GW (electric) each. This amount exceeds the roughly 450 GW of total installed U.S. electric capacity in 1973, of which about 14 GW was nuclear (27 plants) [37]. Obviously, using nuclear power to fuel any major aspect of the hydrogen economy would involve a massive undertaking.

The costs of nuclear power plants are a subject of some controversy, largely because much of the discussion is imprecise in stating (by year) the value of the dollar employed. Moreover, because a nuclear power plant now takes about nine years to construct, the expenditures are not all paid with dollars of a single year's value. Consequently, a discussion of nuclear power plant costs is useful background for this report.

The direct capital costs of a completed plant include the price increases that occur during construction. This price escalation may add 50 percent or more to the estimated direct cost made in terms of dollars valued at the project's starting date. To illustrate, a nuclear plant cost estimate of $300 per kW for equipment, materials, and labor in 1973 dollars may actually incur $450 per kW for direct costs by the time construction is completed seven to ten years later [38]. Interest costs incurred during construction also add to the total installed cost.

TABLE 16.3

Estimated Number of Nuclear/Electrolysis Plants Needed for Select Sectors*

(1973 levels of demand)

| Sector | Number of Plants (1-GW electric, 0.53-GW $H_2$ |
|--------|-----------------------------------------------|
| Aviation | 40 |
| Residential gas | 340 |
| Automotive | 270 |
| Total (for these sectors only) | 650 |

*See Table 12.1 for building-block sizes.

TABLE 16.4

Capital Cost Estimate for Nuclear Plant[a]

|                                          | Dollars per kW |
|------------------------------------------|----------------|
| Direct costs (1973 $)                    | 306            |
| Escalation[b]                            | 142            |
| Allowance for funds during construction  | 114            |
| Subtotal                                 | 562            |
| Use and sales tax                        | 15             |
| Utility costs                            | 41             |
| Contingency (15 percent)                 | 84             |
| Total                                    | 702            |

[a]No land or fuel costs are included.

[b]Equipment, materials, and labor are escalated at different rates.

Source: Reference 38.

Since a utility has to finance construction costs over the construction period, the cost of such funds is properly charged to the capital cost of the asset.

An excellent illustration of the capital cost estimates of nuclear power is contained in a study prepared by Arthur D. Little, Inc., for Northeast Utilities [38]. As shown in Table 16.4, this estimate explicitly breaks out cost escalation and interest during construction. In this case the total funds required for the investment are more than double the 1973 cost of labor, materials, and equipment.

Even with generous allowances for contingencies, most nuclear generating plants do not come on line by the scheduled completion date. The Atomic Industrial Forum (AIF) found that 70 of 95 plants under construction or awaiting construction permits as of December 1973 had experienced delays of 2 months to 5.5 years [39]. Delays were caused primarily by design changes, and secondarily by changes in regulatory requirements and procedures. A third delaying factor was related to labor—either poor productivity or shortages of construction workers.

Delays add significantly to the eventual plant cost, because interest on funds invested before the delay occurs continue to accrue.

Moreover, delays postpone the beginning of revenues, further aggravating the financial situation because utilities must treat interest incurred during construction as a capital investment rather than as expense. From Table 16.4 it can be seen that if there were no escalation of costs during construction, the power plant would cost $560 per kW.* Thus, the total investment (at book value in constant dollars) implied by just the demand sectors using nuclear power to make hydrogen listed in Table 16.3 is about $360 billion (1973 dollars)

As has become apparent throughout this report, the clean burning attribute of hydrogen means that a hydrogen economy would produce less air pollution at the point of end-use than the present system. Indeed, the environment would probably be generally improved in essentially every respect if hydrogen were in widespread use. The major negative environmental aspects of a hydrogen economy occur where the hydrogen is produced; and, as has been seen, unless the coal gasification or solar options develop more vigorously,† this essentially translates into the negative environmental aspects of nuclear power. These aspects have been widely discussed elsewhere and include:

1. Discharge of heat into the air or water (depending on the approach to cooling)
2. Low-level release of radionuclides during normal plant operation
3. Potential accidental release of large amounts of radionuclides during fuel reprocessing and waste disposal

In addition, fundamental questions have been raised about the ability of society to safeguard the plutonium recovered from spent nuclear fission fuel to prevent its falling into the hands of criminals and terrorists [40].

Public debate over nuclear power has been going on almost since World War II ended, and rather than dying down, the level of debate seems to be intensifying—not just over the present kinds of reactors but especially over the breeder reactor, which would produce much more plutonium. Although this report cannot become a treatise on the nuclear power controversy,** it is the crux of the concept of a hydrogen economy: Probably the single most critical factor in the long-term viability of a hydrogen economy is the fate of nuclear power .

---

*Obtained by subtracting $142/kW from $702/kW in Table 16.4.
†Nuclear fusion is also a possibility, although scientific—let alone engineering and economic—feasibility has yet to be demonstrated.
**A bibliography on the topic of nuclear power and its problems is presented at the end of this chapter.

TABLE 16.5

Summary of Impacts of Hydrogen Use in Utilities

| Impact Class | Stakeholder | Nature of Effect | Magnitude of Impact (units) |
|---|---|---|---|
| **Gas utilities** | | | |
| Safety | Gas user | Possibly decreased safety owing to properties of hydrogen | Moderate (people) |
| Business | Utility | Continued viability as an enterprise | Major ($, power) |
| Labor | Utility employees | Effects on work owing to properties of hydrogen affecting safety | Minor (people) |
| | | Continued source of employment | Moderate (people, $) |
| Economic | Utility, consumer, regulatory bodies | Increased cost of energy delivered in gaseous form | Moderate ($) |
| **Electric utility** | | | |
| Reliability | Utility, consumer | Improved reliability owing to modular nature of hydrogen load-leveling storage systems | Moderate ($) |
| Economic | Utility, consumer | Reduced cost of load-leveling technologies | Major ($) |
| | | Reduction of other, more costly forms of spinning reserve | Major ($) |
| | utility, regulatory agencies, consumer | New market possibilities for selling electrolytic hydrogen | Moderate ($, power) |
| | | Cross-subsidization | Minor ($) |
| Environment/ safety | Utility, citizens, regulatory agencies | Increased need for nuclear power plants | Very major (people, power) |

278

Without nuclear power the hydrogen economy concept could rely on coal gasification in the short run and solar power in the long run, but the evolution of the hydrogen economy would be greatly impeded if nuclear power development were impeded.

Table 10.5 gives a short schematic summary of the impacts of hydrogen use in energy utilities.

## APPENDIX: FEDERAL POWER COMMISSION PRIORITIES FOR NATURAL GAS CURTAILMENTS

| Category | Use Criteria ($10^3$ SCF per day) | Description |
|---|---|---|
| | Most Vulnerable | |
| 9 | More than 10,000 | Industrial users with interruptible con- |
| 8 | 3,000 to 10,000 | tracts and boilers equipped to use |
| 7 | 1,500 to 3,000 | other fuels |
| 6 | 300 to 1,500 | |
| 5 | More than 3,000 | Industrial users with firm contracts |
| 4 | 1,500 to 3,000 | and with boilers equipped to use other fuels |
| 3 | | All industrial users not specified in categories 4-9 |
| 2 | | Large commercial users; industrial users with firm contracts who use natural gas for a feedstock, process- sing, or plant protection; distribution companies that store gas for peak- season use |
| | Least Vulnerable | |
| 1 | | Residential and small commercial users who consume less than 50,000 SCF per day |

Source: Federal Power Commission and "Industry Braces for a Natural Gas Crisis," Business Week, October 19, 1974, pp. 114-

REFERENCES

1. A. E. Kahn, The Economics of Regulation, vol. I, Principles; and vol. II, Institutional Issues (New York: Wiley, 1970-71).

2. "Utilities: Weak Point in the Energy Future," Business Week, January 20, 1975, pp. 46-54.

3. F. A. Ford, "A Dynamic Model of the United States Electric Industry, 1950-2010," Research Program on Technology and Public Policy, Dartmouth College, Hanover, N.H., prepared for the National Science Foundation, NSF-RANN Grant no. GI-34808X, June 1, 1975.

4. R. A. Fernandes, "Hydrogen Cycle Peak-Sharing for Electric Utilities," Ninth Intersociety Energy Conversion Engineering Conference, 1974, pp. 413-22.

5. J. E. Hass et al., Financing the Energy Industry (Cambridge, Mass.: Ballinger, 1974).

6. D. P. Gregory et al., "A Hydrogen-Energy System," American Gas Association, Alexandria, Va., August 1972.

7. "Draft Environmental Statement for the El Paso Gasification Project, San Juan County, New Mexico," Upper Colorado Region, Bureau of Reclamation, U.S. Department of the Interior, July 16, 1974.

8. Future Requirements Committee, "United States Gas Consumption 1973," supplement to vol. 5 of Future Gas Consumption of the United States (Denver, Colo.: Denver Research Institute, September 1974).

9. "Natural Gas Users Sing a Dirge," Business Week, January 13, 1975, pp. 37-40.

10. "Industry Braces for a Natural Gas Crisis," Business Week, October 19, 1974, pp. 114-17.

11. U.S., Senate, Committee on Commerce, Federal Power Commission Oversight—Natural Gas Curtailment Priorities, 93rd Cong., 2nd sess., June 20, 1974.

12. "Patterns of Energy Consumption in the United States," Stanford Research Institute, Menlo Park, Calif., for the Office of Science and Technology, Executive Office of the President, Washington, D.C., January 1972.

13. Statistical Abstract of the United States, 1974 (Washington, D.C.: Bureau of the Census, U.S. Department of Commerce, July 1974).

14. "A Furious Push to Deregulate Gas," Business Week, May 9, 1975, pp. 91-92.

15. "Battle over Arctic Gas," Time, April 1, 1974, pp. 20-23.

16. Raymond Huse, Public Service Electric and Gas, Newark, personal communication, 1974.

17. D. P. Gregory, "The Hydrogen Economy," Scientific American, January 1973, pp. 13-21.

18. W. E. Winsche, K. C. Hoffman, and F. J. Salzano, "Hydrogen: Its Future Role in the Nation's Energy Economy," Science 180 (June 29, 1973): 1325-32.

19. E. Fein, "A Hydrogen Based Energy Economy," The Futures Group, Glastonbury, Conn., October 1972.

20. L. T. Blank et al., "A Hydrogen Energy Carrier," Systems Design Institute, National Aeronautics and Space Administration, Johnson Space Center, Houston, Texas, 1973.

21. T. H. Maugh II, "Fuel Cells: Dispersed Generation of Electricity," Science 178 (December 22, 1972): 1273-74B.

22. "Fuel Cell Research Finally Paying Off," Chemical and Engineering News, January 7, 1974, pp. 31-32.

23. D. P. Gregory, "A New Concept in Energy Transmission," Public Utilities Fortnightly, February 3, 1972, pp. 3-11.

24. P. J. Hampson et al., "Will Hydrogen Replace Electricity?" in Hydrogen Energy Fundamentals: A Symposium-Course, ed. T. N. Veziroglu, presented at University of Miami, Coral Gables, Fla., March 1975, pp. S3-25-44.

25. J. M. Burger, "An Energy Utility Company's View of Hydrogen Energy," in Hydrogen Energy Fundamentals: A Symposium-Course, ed. T. N. Veziroglu, presented at University of Miami, Coral Gables, Fla., March 1975, pp. S4-39-63.

26. J. Burger et al., "Energy Storage for Utilities via Hydrogen Systems," Ninth Intersociety Energy Conversion Engineering Conference, 1974, pp. 428-34.

27. P. A. Lewis, "Hydrogen Use by Energy Utilities," in Proceedings of the Cornell International Symposium and Workshop on the Hydrogen Economy, ed. S. Linke (Ithaca, N.Y.: Cornell University, April 1975), pp. 266-73.

28. J. M. Burger, "An Energy Utility Company's View of Hydrogen Energy," in Hydrogen Energy Fundamentals: A Symposium-Course, ed. T. N. Veziroglu, presented at University of Miami, Coral Gables, Fla., March 1975, pp. S3-25-44.

29. S. Law, "Cost of Off-Peak Power," in Proceedings of the Cornell International Symposium and Workshop on the Hydrogen Economy, ed. S. Linke (Ithaca, N.Y.: Cornell University, April 1975), pp. 212-13.

30. G. Strickland et al., "An Engineering-Scale Energy Storage Reservoir of Iron-Titanium Hydride," in Hydrogen Energy, ed. T. N. Veziroglu (New York: Plenum, 1975), pp. 611-20.

31. A. L. Robinson, "Energy Storage (I): Using Electricity More Efficiently," Science 184, no. 4138 (May 17, 1974): 785-87; pt II in Science 184, no. 4139 (May 24, 1974): 884-87.

32. D. A. Brobst and W. P. Pratt, eds., United States Mineral Resources, U.S. Geological Survey Professional Paper 820, U.S. Department of the Interior, 1973.

33. M. Lotker, E. Fein, and F. Salzano, "The Hydrogen Economy—A Utility Perspective," paper presented at the IEEE Winter Meeting, 1973.

34. M. Lotker, "Hydrogen for the Electric Utilities—Long Range Possibilities," Ninth Intersociety Energy Conversion Engineering Conference, 1974, pp. 423-27.

35. "A Concerted Push for Rate Reform," Business Week, April 14, 1975, p. 66.

36. S. Law, "Electric Utility Views," in Proceedings of the Cornell International Symposium and Workshop on the Hydrogen Economy, ed. S. Linke (Ithaca, N.Y.: Cornell University, April 1975), pp. 355-58.

37. "53rd Semi-Annual Electric Power Survey," Electric Power Survey Committee of the Edison Electric Institute, New York, April 1973.

38. "A Study of Base Load Alternatives for the Northeast Utilities System," Arthur D. Little, Inc., Cambridge, Mass., for the Board of Trustees of Northeast Utilities, July 5, 1973.

39. Nuclear News, July 1974, pp. 55-57.

40. M. Willrich and T. B. Taylor, Nuclear Theft: Risks and Safeguards (Cambridge, Mass.: Ballinger, 1974).


## BIBLIOGRAPHY ON NUCLEAR POWER:
## PRO AND CON


### General

Bethe, H. A. "Advanced Nuclear Power." In Report of the Cornell Workshops on the Major Issues of a National Energy Research and Development Program, rev. ed. Ithaca, N.Y.: Cornell University, December 1973, pp. 169-219.

Gillette, R. "Nuclear Power: Hard Times and a Questioning Congress." Science 187 (March 21, 1975): 1058-62.

Hammond, A. "Fission: The Pro's and Con's of Nuclear Power." Science 178, no. 4057 (October 13, 1972): 147-49.

Rose, D. J. "Nuclear Eclectic [sic] Power." Science 184, no. 4134 (April 19, 1974): 351-59.

Weinberg, A. M. "The Short-Term Nuclear Option." In Report of the Cornell Workshops on the Major Issues of a National Energy Research and Development Program, rev. ed. Ithaca, N.Y.: Cornell University, December 1973, pp. 131-67.

_____. "Social Institutions and Nuclear Energy." Science 177, no. 4043 (July 7, 1972): 27-34.

"U.S. Energy Prospects: An Engineering Viewpoint." Task Force on Energy of the National Academy of Engineering, Washington, D.C., 1974.


Nuclear Fuels

Metz, W. D. "Uranium Enrichment: Laser Methods Nearing Full-Scale Test." Science 185 (August 16, 1974): 602-03.

_____. "Uranium Enrichment: U.S. 'One Ups' European Centrifuge Effort." Science 183 (March 29, 1974): 1270-72.

Mohrhauer, H. "Enriching Europe with the Gas Centrifuge." New Scientist, October 5, 1972, pp. 12-14.


Safety and Waste Disposal

"The Deadly Dilemma of Nuclear Wastes." Business Week, March 3, 1975, pp. 70-71.

Gillette, R. "Nuclear Safety: Calculating the Odds of Disaster." Science 185 (September 6, 1974): 838-39.

_____. "Nuclear Safety: Damaged Fuel Ignites a New Debate in AEC." Science 177 (July 28, 1974): 330-31.

_____. "Plutonium (I): Questions of Health in a New Industry." Science 185 (September 20, 1974): 1027-32. "Plutonium (II): Watching and Waiting for Adverse Effects." Science 185 (September 27, 1974): 1140-43.

Kubo, A. S., and D. J. Rose. "Disposal of Nuclear Wastes." Science 182 (December 21, 1973): 1205-11.

## International Politics and Nuclear Theft Risks

McPhee, J.  The Curve of Binding Energy.  New York:  Farrar,
Straus, and Giroux, 1974.

Willrich, M.  Global Politics of Nuclear Energy.  New York:  Praeger,
1971.

_____, and T. B. Taylor.  Nuclear Theft:  Risks and Safeguards.
Cambridge, Mass.:  Ballinger, 1974.


## Breeder Reactor

Friedlander, G. D.  "The Fast-Breeder Reactor:  When, Where,
Why, and How?"  IEEE Spectrum, February 1974, pp. 85-89.

"Second Thoughts That Threaten the Breeder."  Business Week, August 24, 1974, pp. 21-24.

"Suddenly the Gas-Cooled Breeder Looks Good."  Business Week,
February 17, 1975, pp. 36B-36F.

Tinker, J.  "Breeders:  Risks Man Dare Not Run."  New Scientist,
March 1, 1973, pp. 473-76.

"Why the Breeder Program Is under Attack."  Business Week, November, 10, 1973, pp. 222-24.


## Fusion

Metz, W. D.  "Laser Fusion:  One Milepost Passed—Millions More
to Go."  Science 186 (December 27, 1974): 1193-95.

_____.  "Nuclear Fusion:  The Next Big Step Will Be a Tokamak."
Science 187, no. 4175 (February 7, 1975): 421-23.

Post, R. F., and F. L. Ribe.  "Fusion Reactors as Future Energy
Sources."  Science 186 (November 1, 1974): 397-407.

energy budget [2]. Steelmaking consumed about 12 percent of all the world's energy in 1973 [3]. Steel is clearly a commodity of world-wide importance, not only because it is the basic material of industrialized society, but also because of the large proportion of world energy devoted to its manufacture.

In the United States, steelmaking and coal have long been associated industries. In fact, the simultaneous availability of both iron ore and coal has been the dominant factor in plant location. Iron ore is mined primarily in the region around Lake Superior, the largest of the Great Lakes. This ore is concentrated near the mine and then transported by special cargo ships on the Great Lakes to iron-refining and steel-producing regions. Historically, coal has been mined primarily in the Appalachian states, and this had led steelmakers to concentrate in Pennsylvania, Ohio, and Indiana. Table 17.1 shows the dominance of these regions in the industry. Basic steel production employed over 600,000 people in 1973 [4].

The linkage between steel and coal is strong, because the steel industry uses coal both as a source of energy and as a chemical. Steelmaking has been consuming about 20 percent of all U.S. coal production [1],* and conversely coal provides more than 60 percent of the steel industry's energy needs [5]. However, the chemical use of coal (transformed to coke, a hard, porous form of carbon) dominates the energy use, as shown in Table 17.2.

## Steelmaking Processes

Iron ore is an iron oxide.† The iron ore is usually concentrated magnetically and by physical separation processes before shipment to enrich its iron content. At the steel mill, the concentrated ore is then chemically "reduced" to remove the oxygen from the iron oxide to yield unrefined iron. In conventional ironmaking, this reduction is accomplished in a blast furnace. The blast furnace is charged with iron ore concentrate, coke, and a "flux" of limestone or other materials, and then a stream of air is "blasted" through. The coke attracts the oxygen away from the iron oxide ore, and the flux dissolves "gangue" or impurities (mostly silica or alumina) to form "slag," a waste product [1, 6]. The product of a blast furnace is raw or "pig" iron. Thereafter the pig iron is remelted, refined, and made into

---

*This percentage will fall in the future as coal is used in greater amounts to generate electric power.

†There are three possible oxides: ferrous, $FeO$; ferric, $Fe_2O_3$; ferro-soferric, $Fe_3O_4$.

TABLE 17.1

Leading Regions or States Producing Iron Ore, Bituminous Coal,
and Raw Steel, 1972

| Region | Quantity (millions of tons) |
|---|---|
| Iron ore | |
| Total U.S. production | 75 |
| Lake Superior | 62 |
| Western | 11 |
| Total these regions | 73 |
| Bituminous coal | |
| Total U.S. production | 595 |
| West Virginia | 124 |
| Kentucky | 121 |
| Pennsylvania | 76 |
| Illinois | 66 |
| Ohio | 51 |
| Total these states | 438 |
| Steel* | |
| Total U.S. production | 91 |
| Pennsylvania | 20 |
| Ohio | 16 |
| Indiana | 15 |
| Total these states | 51 |

*As indicated by "pig" iron.

Source: Statistical Abstract of the United States, 1974.

steel by controlled additions of small quantities of other materials.
This is accomplished in either an open-hearth furnace, a basic oxygen furnace, or an electric arc furnace. In modern practice the open-
hearth approach is giving way to the other two processes.

The possible future role of hydrogen in steelmaking lies, not
in its use as a fuel, but in its use as a reducing chemical reagent to
replace coke (Chapter 7). The reduction of iron ore without using coke
is termed "direct reduction," and most attention has been given to
processes that utilize a mixture of carbon monoxide (CO) and hydrogen

rather than pure hydrogen as the reducing gas. This mixture is obtained from the reforming of hydrocarbons in the manner discussed in Chapter 4; methane is the preferred feedstock, even though supplies are falling.

Direct reduction of iron using carbon monoxide and hydrogen takes place in the solid rather than the molten state and results in a pelltized or sponge product (95 percent iron) that can substitute for pig iron [1, 7]. The directly reduced iron differs from pig iron, however, because it contains more residual impurities. Use of an electric arc furnace is the preferred way to refine directly reduced iron because it can substitute for the usual charge of steel scrap in the crucible.

Figure 17.1 shows the various present steel production methods in the United States and compares them with the possible future direct reduction/electric arc furnace approach. Although the situation shown in Figure 17.1 represents direct reduction with a carbon monoxide and hydrogen mixture, reduction by pure hydrogen would involve the same sequence.

A special study committee of the American Iron and Steel Institute (AISI) concluded, however, that use of pure hydrogen is less desirable than use of the gas mixture, because with pure hydrogen there is a tendency for the iron particles to sinter (stick together) and to reoxidize [6]. Nevertheless, if hydrogen were used, these processing problems could almost certainly be overcome.

There is now considerable discussion concerning the use of nuclear energy in steelmaking. First, nuclear heat would be used to reform hydrocarbons to obtain a reducing gas. Second, electricity

TABLE 17.2

Uses of Coal in the Steel Industry, 1970

| Use | Quantity ($10^6$ tons) | Percentage of Total* |
|---|---|---|
| Coke making | 87.0 | 95.0 |
| Steam production | 4.8 | 5.0 |
| Miscellaneous | 0.3 | 0.3 |
| Total | 92.0 | 100.0 |

*Percentages do not add because of rounding.

Source: Reference 1.

FIGURE 17.1

Comparison of Conventional Steelmaking and Direct Reduction
Approach

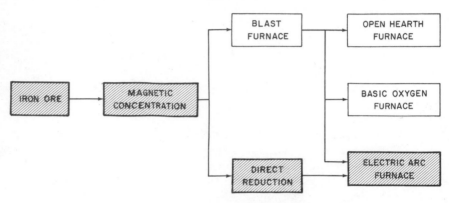

Note: Entire direct reduction option shown shaded.

would be generated for the electric furnace [2, 6, 8]. If pure hydrogen were used, then nuclear energy would be used to produce it either
electrolytically or by closed thermochemical cycles. Table 12.1
showed that there is a good match between a nuclear-electrolytic plant
building block and a direct reduction building block if about half the
nuclear plant's output is used as electricity and half is used to make
hydrogen.

Thus, once again, the question of society's acceptance of nuclear
power becomes critical in this aspect of the hydrogen economy. The
question is more difficult here, however, because the only U.S.-built
nuclear reactor presently capable of supplying the requisite high temperature (1,650° F) is the high-temperature, gas-cooled reactor (HTGR)
made by General Atomic Company—a reactor that requires weapons-
grade fuel [9]. This fuel greatly exacerbates the danger of theft of
nuclear materials by terrorist groups.

Most current thinking about direct reduction processes continues
to assume the use of reformed methane. Consequently, much interest
has been shown by foreign countries that have vast surpluses of methane—especially the Soviet Union and countries in the Middle East.
Other countries also showing interest in direct reduction are those not
so well endowed with natural resources as the United States. Japan is
said to be especially concerned about the reduced availability of coals
suitable for coking (metallurgical coal) and is quite interested in the
rect reduction process [2, 3, 6]. Yet, because the United States is

still well endowed with all the traditional steelmaking ingredients (iron ore, coking, coal, and limestone), the U.S. steel industry is not pursuing direct reduction with vigor. The AISI nevertheless concluded that the potential for direct reduction warranted expanded research and development efforts in this country [6].

## Impacts

The major consequences of a switch to direct reduction techniques would be the following [1]:

1. Reduced need for coke (about 50 percent) and, therefore, in proportion, metallurgical coal
2. Great reduction in air pollution because of the reduced production of coke
3. Reduced production of slag and, therefore, in proportion, need for flux materials, such as limestone
4. Great reduction in the amounts of oxygen consumed in steelmaking as the electric furnace substituted for basic oxygen furnaces
5. Increased productivity (about 45 percent), since the time needed for an electric furnace cycle would be reduced from 160 minutes to 110 minutes
6. Reduced need for steel scrap in steelmaking
7. Reduced capital investment in steelmaking
8. Accentuated desirability of ores with low levels of impurities

As with most technological changes, however, the most convincing driving force is economic savings.

The AISI committee on direct reduction estimated that the conventional (nonnuclear methane-reforming approach to the direct reduction/electric furnace method would require a capital investment of $80 per annual-ton compared with an investment of $130 per annual-ton for the conventional blast furnace/basic oxygen furnace method. There would also be a reduced selling price of steel. However, if the heat for the methane-reforming process were supplied by HTGR, the investment requirement would be about $140 per annual-ton; this is about 8 percent higher than the conventional approach employing blast furnaces [6]. The conclusion that nuclear steelmaking is not yet economically competitive is apparently contradicted, however, by a British study that concluded that nuclear steelmaking would already be cheaper than the conventional approach [2].*

---

*Because of a lack of data, we are unable to reconcile these claims.

A key advantage of direct reduction is the elimination of much coke production, because this process emits large quantities of air pollutants. Indeed, air pollution from coke ovens is, without doubt, the steel industry's major current environmental problem [5, 6, 10, 11]. Clearly, the need for environmentally disruptive coal mining is correspondingly reduced. These environmental benefits, however, are somewhat offset by the need to concentrate the raw ore more completely near the mine. This would result in increased solid waste for disposal in the mining region. However, increased concentration reduces the need for impurity-removing, slag-producing fluxes, such as limestone, thereby again reducing mining activity and also reducing slag production at the smelter. Since slag is often recycled and completely consumed as a construction material (road aggregate, railroad ballast, and so on), a reduction in slag is less beneficial than it would seem at first glance [1]. Table 17.3 summarizes the changes in quantities of inputs and outputs for carbon monoxide-hydrogen direct reduction processes. Use of pure hydrogen for direct reduction would result in a similar table.

An important consequence of direct reduction processes with environmental implications is its effect on the recycling of scrap iron and steel. The ferrous scrap market is very volatile [1], and business success in this field is elusive. Because direct reduction of iron can substitute for the scrap normally used in electric furnaces, direct reduction might deal the domestic market for ferrous scrap a crushing blow [1]. Since steelmaking consumes about 5 percent of all U.S. energy, ferrous scrap should be viewed not only as a materials resource but also as a stored energy resource. In an energy-tight future, this resource should be used diligently to conserve energy and to reduce the environmental effects of energy production and consumption. Hence, a reduction in the economic viability of the ferrous scrap recycling business must be viewed as detrimental.

Where direct reduction steelmaking is being practiced, small steel mills compete effectively with large mills using conventional practices [11]. Thus, an important impact of a transition to direct reduction in steelmaking would be the increased viability of so-called mini-mills. Even without direct reduction, mini-mills are already beginning to spread the production of steel around the country and to lessen the former geographical concentration [1, 11]. Most mini-mills are said to be locating near the coastlines to facilitate both the import of foreign ore concentrates and water shipment of finished products [1]. This dispersal trend could have profound social implications, because over 600,000 workers are directly engaged in the basic steel industry and most are concentrated in three states—Pennsylvania, Ohio, and Indiana. Over the long term, as nuclear powered/direct reduction steelmaking becomes increasingly viable, a signifi-

## TABLE 17.3

Summary of Changes Expected for Direct Reduction/Electric
Furnace Steelmaking[a]

| Inputs and Outputs | Percentage Change Compared to Average U.S. Practice in 1970 |
|---|---|
| Ferrous inputs | |
| Ore | −53 |
| Scrap | 0[b] |
| Fuels and energy | |
| Coke (coal) | −77 |
| Purchased electric power | +310 |
| Natural gas | +1,100[a] |
| Fuel oil | −100 |
| Fluxes | |
| Limestone and dolomite | −62 |
| Lime | +34 |
| Fluorspar | −40 |
| Oxygen | −51 |
| Slag produced | −60 |

[a]Assumes conventional reforming of methane, use of carbon monoxide–hydrogen mixture as reducing gas; no use of nuclear power.

[b]Assumes no change, although there is the possibility of displacing much scrap.

Source: Reference 1.

cant fraction of the steel industry may relocate as it optimizes a new set of parameters in which the location of available metallurgical coal is much less important than previously.

### Steel Summary

Although the U.S. steel industry is likely to embrace nuclear powered/direct reduction steelmaking less quickly than other countries because of the favorable U.S. resource position, a transition would have important environmental benefits (Table 17.4). Especially important would be the reduction in air pollution. Potentially profound

social consequences would appear as relaxed locational constraints
resulting in a new geographical distribution of the steel industry.


## AMMONIA SYNTHESIS

Ammonia synthesis is the largest single use of hydrogen today.
As discussed in Chapter 7, this hydrogen is obtained from the chemical
reforming of methane.  Although methane is becoming increasingly
scarce and expensive (when not subject to federal price regulations),
the priority held by ammonia producers to supplies of natural gas is
second only to the priority of residential use [12, 13].  This priority
is a direct result of the fact that about three-fourths of all ammonia
is being used as an agricultural fertilizer.

Ammonia synthesis is a relatively simple and extremely well
developed chemical engineering process.  Modern plants use their
methane feedstock very efficiently to obtain hydrogen as a fuel and,
indirectly, to separate nitgoren from the air (Chapter 7).  If ammonia
producers were unable to obtain their hydrogen from methane they
could obtain it from other hydrocarbon sources more cheaply than from
electrolytic hydrogen.  Although the average interstate (regulated)
price of natural gas was about $0.22/10^3$ SCF in early 1974, some recent prices of unregulated intrastate gas reached $2.00/10^3$ SCF in
mid-1975 [14].  It is reasonable to expect unregulated intrastate natural gas prices to level off at about $2.50/10^3$ SCF before 1980 [16].

Table 17.5 compares the cost of alternative feedstocks that yield
hydrogen at the same cost to an ammonia producer.  It can be seen
from Table 17.5 that electrolysis of water using electricity at $0.01
per kWh (a realistic cost) could not produce hydrogen competitively
with reforming of either natural gas or other hydrocarbons until
those fossil fuels had risen in cost dramatically.  Moreover, it must
be recalled that a rise in cost of those fuels would, in turn, cause the
cost of electricity to rise by virtue of competition between energy forms
and the need to use primary energy to generate electricity.

It is apparent, therefore, that a considerable rearrangement
would have to occur in both allocation priorities for natural gas and
the relative prices of various energy forms before hydrogen from coal
or electrolysis could be expected to become attractive to ammonia producers.  Moreover, as other lower-priority users of natural gas are
curtailed, the lifetime of the supplies will be extended, which will increase the likelihood that ammonia producers will continue to receive
methane on a priority basis.

If access to methane were to change, however, the building-
block description of Table 12.1 shows that a single nuclear-electrolytic plant could supply almost exactly two ammonia synthesis building

# TABLE 17.4

## Summary of Impacts of Nuclear/Direct Reduction/Electric Furnace Steelmaking

| Impact Class | Stakeholder | Nature of Effect | Magnitude of Impact (units) |
|---|---|---|---|
| **Environment** | | | |
| Air pollution | Steel industry | Reduced air pollution from coke making | Major (people, $) |
| | EPA | | |
| | Public | | |
| Mining | Steel industry | Reduced mining for coal (but demand slack taken up by other energy needs) | Minor ($, people) |
| | Coal industry | | |
| | Limestone industry | Reduced mining for limestone | Moderate ($, people) |
| | Public | | |
| Solid waste | Steel industry | Increased tailings from increased concentration of ore | Moderate (people, $) |
| | EPA | Decreased slag production | Minor ($) |
| | Public | | |
| Resources | Public | Improved utilization of natural resources | Moderate ($) |
| Safety | Public | Increased use of nuclear power, especially with weapons–grade fuel | Major (people) |
| **Social** | | | |
| Industry location | Steel industry | Dispersal of steelmaking from present geographical concen–tration | Major ($, people) |
| | Labor | | |
| Resource extraction | Coal industry | Reduced need for metallurgical–grade coal (but slack taken up by other energy demands) | Minor ($, people) |
| | Labor | | |
| **Economic** | | | |
| Capital investment | Steel industry | Potentially reduced investment per unit output | Minor ($) |
| | Capital market | | |
| | Stockholders | | |
| Production costs/ market price | Steel industry | Potentially reduced unit cost | Minor ($) |
| | Consumers | | |

TABLE 17.5

Estimated Feedstock Prices That Would Produce Hydrogen at Equal
Costs (About $1.50 per $10^3$ SCF) by Various Processes
(in dollars)

| Process | Approximate Feedstock Price (conventional unit) | Energy Unit (per $10^6$ Btu) |
|---|---|---|
| Steam reforming of methane* | 4.70 per $10^3$ SCF | 4.95 |
| Electrolysis of water | 0.010 per kWh | 2.93 |
| Alternative hydrocarbons | | |
| Steam reforming of naphtha | 0.50 per gallon | 4.17 |
| Partial oxidation of naphtha | 0.40 per gallon | 3.16 |
| Partial oxidation of residual oil | 17.00 per barrel | 2.70 |
| Gasification of bituminous coal | 46.00 per ton | 2.78 |

*At this high price for methane feedstock all other contributions
to the cost of steam-reformed hydrogen are insignificant [16].

Source: Stanford Research Institute.

blocks. However, since even this demand falls short (only about one-
third) of the capacity of a 24-inch diameter hydrogen pipeline building
block, the hydrogen production would probably have to be quite close
to the ammonia plant and employ a suboptimum delivery pipeline.
Consequently, analysis of the relevant building blocks suggests that
ammonia synthesis would have to either rely on electrolytic hydrogen
production facilities dedicated to such synthesis or wait until other
large hydrogen demands had already justified deployment of large
delivery pipelines.*

By far the most important consequence of ammonia producers
joining in the hydrogen economy would arise from their changed criteria for plant location. Once access to abundant and low-cost natural
gas ceased to be the key variable, the plants would be free to locate
close to their demand because the key plant feedstocks (water and nitrogen from the air) are widely available. Today most ammonia is shipped

*The reasons discussed here are the underlying justification for
showing a late ammonia synthesis transition in the realistic implementation scenario depicted in Figure 13.2.

to market by combinations of barge, rail, and truck. However, several fairly large-capacity ammonia pipelines are being built to connect producing regions to midwestern farming regions. Consequently, access to these pipelines might continue to constrain plant locations, even though, in principle, the plants could locate anywhere adequate supplies of primary energy and water could be obtained. However, to the extent that this product pipeline constraint was not honored, the ammonia synthesis plants would almost surely locate in the farm regions—especially in the Midwest—to be near the market. This implies a dispersal of the industry and an effect on the workers employed in production and transportation.

## REFERENCES

1. R. J. Leary and G. M. Larwood, "Effects of Direct Reduction upon Mineral Supply Requirements for Iron and Steel Production," Bureau of Mines Information Circular 8583, 1973.

2. N. Valery, "Steelmaking with Heat from the Atom," New Scientist, September 13, 1973, pp. 610–15.

3. "The New Economics of World Steelmaking," Business Week, August 3, 1974, pp. 34–39.

4. Statistical Abstract of the United States, 1974 (Washington, D.C.: Bureau of the Census, U.S. Department of Commerce, July 1974).

5. "A Search for Clean Coking Processes," Steel Facts, no. 1, American Iron and Steel Institute, Washington, D.C., 1974.

6. D. J. Blickwede and T. F. Barnhardt, "The Use of Nuclear Energy in Steelmaking," paper presented at the First National Topical Meeting on Nuclear Process Heat Applications, Los Alamos Scientific Laboratory, Los Alamos, N. Mex., October 1–3, 1974.

7. E. Fein, "A Hydrogen Based Energy Economy," The Futures Group, Glastonbury, Conn., October 1972.

8. "Nuclear Reactors for Steelmaking," Business Week, October 19, 1974, p. 52P.

9. M. Willrich and T. B. Taylor, Nuclear Theft, Risks and Safeguards (Cambridge, Mass.: Ballinger, 1974), p. 43.

10. "Coke Oven Control Program Could Cost Bethlehem Steel $40 Million," Air/Water Pollution Report, October 28, 1974, p. 428.

11. "Hydrogen: Likely Fuel of the Future," Chemical and Engineering News, June 26, 1972, pp. 14–17; "Hydrogen Fuel Use Calls for New Source," Chemical and Engineering News, July 3, 1974, pp. 16–18; and "Hydrogen Fuel Economy: Wide Ranging Changes," Chemical and Engineering News, July 10, 1972, pp. 27–29.

12.  "Natural Gas Users Sing a Dirge," Business Week, January 13, 1975, pp. 37-40.

13.  "Pushing for Priority," Chemical Week, April 3, 1974, p. 15.

14.  "A Furious Push to Deregulate Gas," Business Week, May 19, 1975, pp. 91-92.

15.  Robert Muller, Process Evaluation Department, Stanford Research Institute, Menlo Park, Calif., personal communication, 1974.

16.  "Hydrogen and Other Synthetic Fuels," a summary of the work of the Synthetic Fuels Panel, prepared for the Federal Council on Science and Technology R&D Goals Study, September 1972.

air pollutant emissions:
automotive, 189-91, 197,
239; aviation, 225, 237-
39, 245, 246; power-
plant, 265; in steelmaking,
291, 292, 293
Air Products and Chemicals,
Inc., 154, 155
Allis-Chalmers, 146, 151
American Automobile Asso-
ciation, 185
American Iron and Steel Institute (AISI), 289, 291
ammonia, synthesis of, 96-
97, 166, 170-71, 286,
294-97
Arab oil embargo, 220
Atomic Industrial Forum
(AIF), 276
Atoms-for-Peace program,
134
automobiles: liquid hydrogen
fuel for, 85-86, 130, 158,
198, 200; metal hydride
fuel for, 58, 69-70, 85,
86-88, 130, 198, 200-01;
superior efficiency of
electric cars, 203
automotive fuel, hydrogen as,
84-88, 130-35, 175, 183-
209; costs of fuel systems,
200-01; decrease in air
pollutants, 189-91, 197;
filling-station problems,
198-99; for fleet use, 88,
132-34, 205-09; gasoline-
based system, comparative

value of, 201-03; gasoline-
based system, disinterest
of, 183-87; nonair pollution,
probable increase in, 191-
92; other vehicles, 88, 134-
35; in private cars, 84-88,
130-32, 175, 184-205; re-
duced dependence on for-
eign fuels, 197-98; safety
factors involved, 194-95;
special training needed to
handle, 195-97 [less owner
repair, 196-97]; strategies
of transition to, 188-89
aviation, hydrogen as fuel for,
4, 6, 69, 89-91, 135, 175,
179; design changes re-
quired, 89-91, 222-23;
government role in deci-
sion for, 247-50; transition
to, 227-32 [advance re-
search needed, 232; com-
parative cost against synthetic jet fuel, 228-29;
concerns of stakeholders
in, 229-31; enthusiasm
for, 231-32; fewer decision
makers, 227; fine coordination needed, 227-28; hypothetical schedule for, 228];
value of liquid, 89, 135

Bethlehem Steel Corp., 99
Boeing Aircraft, 235
Bonneville Power Administra-
tion, 253

Brookhaven National Laboratory, 61, 82, 124

CAB (Civil Aeronautics Board), 221, 225-27, 233
catalytic burners, flameless, 73, 75-76, 129-30
catalytic converters, 185
catalysts, 42, 124
Chao, R. E. and K. E. Cox, 148
Chemical Engineering, 150
chemical uses, of hydrogen, 77-78, 95-99, 288-94; ammonia synthesis, 96-97, 294-97; coal gasification or liquefaction, 97; reduction, in steelmaking, 98-99, 288-94
coal, 4, 7, 20, 37, 97, 121, 122, 159, 241, 243, 246, 254, 274, 277, 279, 287-88, 290; coking, 98, 290, 291, 293; gasification of, 20, 37, 97, 129, 144-45, 160, 166, 256, 258; lignite, 122
commercial aviation, U.S.: airlines, competition among, 214-16; airport operators, 219 [fuel contracts with, 219; and noise abatement, 219, 239]; chief stakeholders in, 213; development of new planes, 217-19; few top aviation hubs, 214; future fuel choice, need for, 220-21; government regulatory agencies, role of, 221-27 [CAB, 225-27, 231; EPA, 224-25, 231; FAA, 221,

223-24, 231; NTSB, 222]; and military experience, 222, 235, 250; operational savings desired by, 216-17
costs, of hydrogen technologies, 140-61; comparative production costs, 150-51; comparative systems costs, 158-60, 171 [government policies in, 160]; electrolysis of water, 146-47; gas storage, 156-57; gasification from coal, 144-45; liquefaction, 154-55; storage as liquid, 157; thermochemical processes, 147-49; transport as liquid, 155; transport of gas, 151-54
Cox, K. E. (see, Chao, R. E. and K. E. Cox)

deBeni and Marchetti, 148
Defense, Department of (DoD), U.S., 218-19, 222, 241
dewars, 54, 56, 200, 201
direct reduction, of iron ore, 288-89
Douglas Aircraft Co., 235

El Paso Natural Gas Company, 256
electric utilities, 78-84, 264-74; with gas combination, advantages of, 84, 273-74; hydrogen production and sale by, 270-73; hydrogen use unlikely in power generating, 258-60; hydrogen/fuel cell storage, 80, 265, 267, 269; interdependence

of, 264; load factor, 78–
79; load leveling, 78–80,
130, 194, 265, 267–70,
272, 273, 274; long-
distance energy transmis-
sion, 82–84, 123, 124,
265–67; major problems
of, 265; natural gas use
by, 257–58; nuclear power,
difficulties of, 82, 265;
pumped hydroelectric
storage, 80, 127; spinning
reserve, 268, 269, 274;
stakeholders in, 264; technologies available to,
122–23
electricity, 3, 4, 6, 7, 8, 21,
23, 122, 130, 146, 260,
294; heat pumps using, 130
electrolysis, 3, 24, 38–42, 44,
47, 124
embrittlement of metals, 61,
65–66
energy, 3, 77, 269, 274; hydrogen's advantages as
carrier, 23–24; options
for meeting demands, 20–22
energy utilities, 73–84, 253–
79; capital-intensive in-
dustry, 254–55; definition
of, 253; hydrogen's role
in, 255; public nature of,
254; rates of, 253–54;
regulatory commissions
for, 253, 271, 272
Environmental Protection
Agency (EPA), 185, 224,
225
Euratom, 43, 148, 151


FAA (Federal Aviation Administration), 219, 221, 222,
223, 224

Federal Aviation Regulations
(FARs), 221, 224, 239
Federal Power Commission
(FPC), 73, 253, 262
fuel cells, 40, 75, 76, 82,
123, 124, 130
Futures Group, 148, 151, 155,
200


gas utilities, 73–78, 255–64;
consumers, 257–58; im-
pacts of hydrogen conver-
sion, 261–64 [consumer
problems, 262–64; costs,
262; parallel systems re-
quired initially, 261–62];
stakeholders in, 255;
structure of industry, 255–
56
General Atomic Company, 43,
290
General Electric Co., 124,
148, 264
Geological Survey, U.S., 270


Hanneman, R. E. (see, Wentorf,
R. H., Jr., and R. E.
Hanneman)
helium, 52, 55, 124–25; liquid,
53, 94, 99, 100, 124
high-temperature, gas–cooled
reactors (HTGR), 290, 291
Hindenburg, 112–13, 245
hydrogen: advantages of, as
fuel, 23; alternatives to,
121–36 [for automotive
fuel, 130–35; for aviation
fuel, 135; for consumer
uses, 129–30; as energy
carrier, 121–24; in energy
distribution, 124–25; for
energy storage, 127–28; as

reducing agent, 135]; as chemical reactant, 77-78, 95-99, 288-97; by closed-cycle thermochemical water decomposition, 42-45, 99, 124, 147-49, 150, 151, 191, 265, 290; cross-link to electricity, 3; cryogenic uses for, 99-100; distribution of, 64-70, 151-55; embrittlement of metals by, 61, 65-66, 100, 113, 124, 132; an energy carrier, 3, 23, 35, 121; from fossil fuels, 35-38, 99, 143, 144-45, 150, 241, 294; in household uses, 74-77, 129-30; hydrogen/electric future, 3, 179; for industrial energy uses, 77, 260-61; liquefaction of, for storage, 55-57; liquid, 13, 51, 53-57, 68-69, 85-86, 88, 89-91, 94, 95, 99, 113, 121, 123, 124, 125, 135, 154-55, 157, 220, 232, 234; long-term transition to, 4-9; military usages impractical, 94-95; pipeline transmission of, 53, 64-68, 69, 70, 113, 151-54, 155, 166-68, 265; possible government policies toward, 5-7, 115-16, 171-75; priorities of research toward use of, 10-14; safety of, 106-16, 194-95, 263-64; sources of, 35; in steelmaking, 98-99, 286, 288-94; storage of, 51-61, 85, 111, 123, 124, 127, 156-57, 268-70; technology assessment for,

24-25; from thermonuclear fusion, 45-47; use of, 3-4; for vehicle fuel, 84-94, 130-35, 158; from water by electrolysis, 38-42, 44, 47, 99, 143, 146-47, 150, 151, 191, 265, 290, 294
hydrogen-fueled commercial aircraft, impacts of transition to, 232-47; economic, 232-37 [airport financing, 233; employment opportunities, 237; foreign sales, 235-37; maintenance time, 232, 234; manufacturers' problems, 235-37; new fueling infrastructure needed, 232-33; routing and scheduling, 234-35; turnaround time, 232, 234; used-aircraft market, 233]; environmental, 237-41 [air pollution, 237-39, 245, 246; noise, 239-41, 244, 245, 246]; resource utilization, 241-43; social, 243-46 [decreased air travel, 243-44; safety, 244-46]; technological improvements, 246-47

Institute of Gas Technology, 151
Internal Revenue Service, 233

KMS Fusion Corp., 45
Koppers-Totzek gasifier, 144-45

lead-free gasoline, 185, 187
Linde Division (see, Union Carbide Corporation, Linde Division)

lithium sulfur batteries, 203,
205
Little, Arthur D., Inc., 110,
276
LNG (see, natural gas, liquid)
Lockheed Aircraft Corp., 89,
235, 238, 239
Los Alamos Scientific Laboratory, 124


magnesium: abundance of,
200-01; costs of, 200
Mark I thermochemical
process, 148
Mark IX thermochemical water
decomposition process, 43
Mazda Wankel (rotary) engine,
185
methane, 10, 13, 35-36, 37,
45, 56, 64-65, 66, 68, 73,
76, 77, 78, 96-97, 99,
106, 108, 109-10, 111,
122, 127, 170, 262, 289,
290, 291, 294; liquid, 134,
135 (see also, natural gas)
methanol, 61, 85, 131, 134,
135
Minnesota Valley Engineering
Company, 200


NASA, 89, 94, 111-12, 222
National Center for Atmospheric
Research (NCAR), 239
National Science Foundation, 17
National Transportation Safety
Board (NTSB), 222
natural building blocks, 165-
66; of hydrogen economy,
166-69, 189, 220, 266-
67, 273, 275, 290, 294-96
natural gas, 3, 5, 7-8, 35, 37,
64-65, 66, 73, 76, 77, 78,
99, 106, 122, 129, 143,
159, 160, 170, 241, 254,
255, 260, 294, 296; cur-
tailment priorities for, 170-
71, 255, 256-57, 258, 294;
liquid (LNG), 7-8, 125, 127
(see also, methane)
networks, resistance to change
by, 183-84
nickel, 42, 200, 201
nitrogen, 61, 294, 296; liquid,
53, 100; oxide, 189, 238,
239
noise, 219, 225, 239-41, 244,
245, 246
Noise Exposure Forecast (NEF),
225, 240
Northeast Utilities, 276
nuclear power, 4, 8, 21, 80,
82, 97, 98, 123, 134, 135,
147-49, 155, 166, 191,
239, 246, 264, 265, 273-
79; cost of plant construction, 275-77; critical importance of, to hydrogen
economy, 277-79; negative
environmental aspects of,
277; number of plants needed
for electric-hydrogen economy, 275


Office of Technology Assessment (OTA), 17, 18
Organization of Petroleum Exporting Countries (OPEC),
159


petroleum fuels, 3, 5, 7-8, 20,
121, 135, 144
platinum, 42

plutonium, 277
Post Office, U.S., 134, 206
Project Independence, 22, 159, 160
Public Service Electric and Gas (PSE&G, New Jersey), 80-82, 271

Research Applied to National Needs (RANN) Program, 17

safety, of hydrogen, 106-16, 194-95, 263-64; in automobile collisions, 195; behavior in spills, 108-09; cold "burns," 113; dispersal, 108-09, 111, 195; flammability, 109-11; hydrogen economy, control of, 115-16; low radiant energy, 195; metals embrittlement, dangers of, 113; in metal hydrides, 111, 195; properties, 106-08; public understanding of, 114-15
Savannah, SS, 134
solar energy, 8, 21, 47, 123, 127, 149, 191, 274, 277, 279
space shuttle, 94
Stanford University, 124
steam, process, production of, 77, 260
steelmaking, 98-99, 286-94; by direct reduction of ore, 288-94 [consequences of, 291; environmental benefits of, 292; viability of mini-mills, 292-93]; energy needed for, 286-87;

linkage of coal to, 287; nuclear energy for, 289-90; processes of, 287-91
storage, of hydrogen, 268-70; in chemical compounds, 61; linepack for gaseous, 51-53, 156-57; as metal hydrides, 57-61, 85, 111, 124; in tanks for gas, 157; in tanks for liquid, 53-55, 85, 111, 123, 127, 157; underground for gaseous, 51-53, 156-57
substitute natural gas (SNG), 37, 122, 129, 159, 170, 256, 261, 264
superconducting, 99, 125
Synthetic Fuels Panel Report, 42, 146
Synthetic Fuels Task Force, 159
synthetic hydrocarbon fuels, 121-22, 186, 221, 229, 235, 241-43

tank ships, for liquid hydrogen transport, 124, 125
technology assessment, 17-18; of hydrogen, 24-25, 120
Teledyne Isotopes, 146
Tennessee Valley Authority, 253
Texas Gas Transmission Co., 45
titanium, 270
transition, to hydrogen economy, 170-79; basic considerations for, 170-71; decision-making climates, 171-75; eras of, 175; far off, 178-79; selected end-uses, importance of, 175-78

Union Carbide Corporation,
    186; Linde Division, 55,
    56, 124, 150, 154, 155,
    157, 159

Wentorf, R. H., Jr. and R. E.
    Hanneman, 148
Westinghouse Electric Corp.
    264